Selected Developments in Catalysis

Critical Reports on Applied Chemistry Volume 12

Selected Developments in Catalysis

Edited by J.R. Jennings

Published for the Society of Chemical Industry by
Blackwell Scientific Publications
Oxford London Edinburgh
Boston Palo Alto Melbourne

© 1985 by Society of Chemical Industry
14–15 Belgrave Square, London, SW1X 8PS
and published for them by
Blackwell Scientific Publications
Osney Mead, Oxford, OX2 0EL
8 John Street, London, WC1N 2ES
23 Ainslie Place, Edinburgh, EH3 6AJ
52 Beacon Street, Boston
    Massachusetts 02108, USA
667 Lytton Avenue, Palo Alto
    California 94301, USA
107 Barry Street, Carlton
    Victoria 3053, Australia

First published 1985

Photoset by Enset (Photosetting),
Midsomer Norton, Bath, Avon
and printed and bound by
Clark Constable, Edinburgh, London, Melbourne

DISTRIBUTORS

USA and Canada
    Blackwell Scientific Publications Inc
    PO Box 50009, Palo Alto
    California 94303

Australia
    Blackwell Scientific Publications
    (Australia) Pty Ltd
    107 Barry Street
    Carlton, Victoria 3053.

British Library
Cataloguing in Publication Data

Selected developments in catalysis.—(Critical
    reports on applied chemistry; v. 12)
    1. Catalysis
    I. Jennings, J.R.   II. Series
    660.2′995   TP156.C35

    ISBN 0 471 91324 3

Contents

Introduction

The acceleration of a chemical reaction by species which themselves are not consumed during that reaction has been observed since the early days of chemistry. The first industrial application of this effect in 1746, was in the manufacture of sulphuric acid by the lead chamber process, using nitric oxide to aid the oxidation of sulphurous to sulphuric acids. Subsequent developments in the early nineteenth century included the discovery of such gas phase reactions as catalytic combustion and the oxidation of sulphur dioxide to sulphur trioxide which in later years was to supercede the lead chamber process. The term 'catalyst' was proposed by Berzelius in 1836 for those substances which could promote (or retard) chemical reactions. Many definitions of catalysts have been offered, but Ostwald's definition, namely that 'a catalyst is a substance that alters the velocity of a chemical reaction without appearing in the products', is still useful. As we now know, catalysis has been with us far longer than the days of the lead chamber process. Biological processes depend extensively upon enzyme catalysis without which life as we know it could not be sustained (and at a more basic level would deprive mankind of all natural alcoholic beverages!).

Since the early days, the development of catalysts and the science of catalysis has gone hand in hand with the development of the chemical industry. Few processes and few products do not depend at least in part on catalysis at some stage in the manufacturing chain. Though many of these products could be prepared by non-catalytic reaction, the costs would be prohibitive by today's standards. Without catalysis, the chemical industry would have evolved differently and we would have been denied such everyday products as polyolefins, most man-made fibres, fertilizers and many of the resins used in the building trade.

Catalysts can operate either homogeneously or heterogeneously. Homogeneous catalysis is mostly performed in the liquid phase with the catalyst dissolved in the reaction medium, as for example in the acid-catalysed hydrolysis of organic esters. In heterogeneous catalysis, the reactants, in mainly the gaseous or liquid phases contact catalytically active surfaces. Catalysis of reactions in the solid state is less common; an example is the manganese dioxide catalysed decomposition of potassium chlorate to oxygen and potassium chloride. The relative dominance of heterogeneous catalysis in industrial practice is reflected in the contents of this volume, in which four of the chapters refer mainly to heterogeneous catalysis.

Most of the catalyst development work has been in the areas of activity, selectivity and stability. Activities have been increased by a variety of different methods. For example, heterogeneous catalytic activity is frequently proportional to surface area and methods which increase the area of the active species have also led to an increase in activity. A good example of this is the ammonia synthesis catalyst developed by Mittasch and co-workers. Dissolution of alumina in magnetite by fusion, gave a product which on reduction consisted of iron crystallites on an alumina skeleton, having not only higher surface areas but also higher activity. The use of promoters has also proved to be successful in the search for higher activities. Potash promotion of ammonia synthesis catalyst is well known. Other examples include the addition of potash to copper-based oxychlorination catalysts, chloride promotion of silver-based epoxidation catalysts and perhaps cobalt promotion of molybdenum hydrodesulphurization catalysts.

The second main area of catalyst development is that of selectivity. Many different compounds may be thermodynamically feasible from a given mixture of starting components. This is amply illustrated by the hydrogenation of carbon monoxide where a range of compounds can be obtained, according to catalyst and conditions, as shown below.

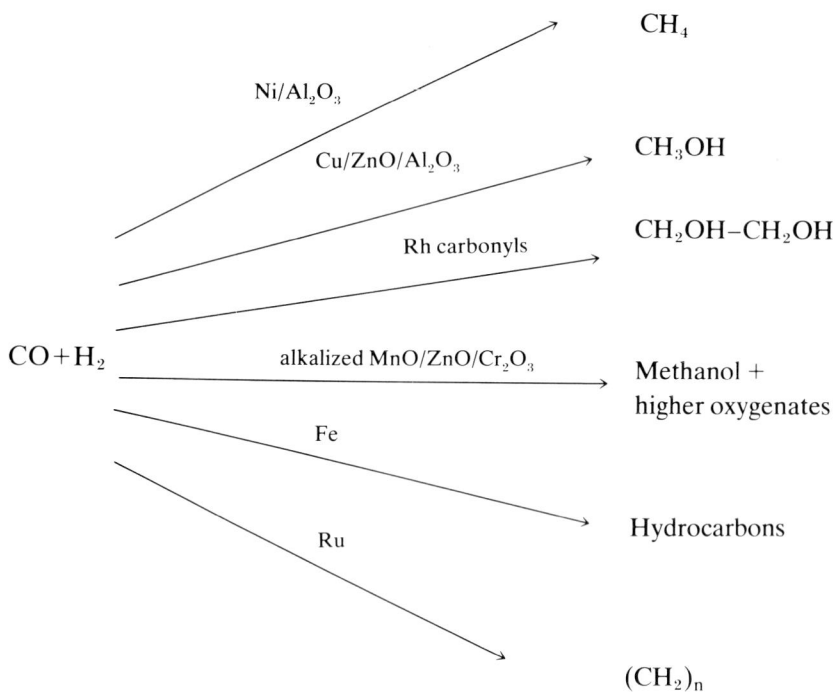

$$CO + H_2$$

Ni/Al$_2$O$_3$ → CH$_4$

Cu/ZnO/Al$_2$O$_3$ → CH$_3$OH

Rh carbonyls → CH$_2$OH–CH$_2$OH

alkalized MnO/ZnO/Cr$_2$O$_3$ → Methanol + higher oxygenates

Fe → Hydrocarbons

Ru → $(CH_2)_n$

The most thermodynamically favourable compound in this range is methane, with methanol the least favourable. Despite this, however, it is possible to synthesize methanol from this mixture with greater than 99% selectivity by judicious choice of catalyst. It is unfortunately often the case that the most active catalysts are the least selective and it may be necessary to sacrifice rate to achieve the required product distribution.

The problem of catalyst stability has major industrial significance. All catalysts deteriorate on line, and the reaction vessel needs to be sized to generate sufficient conversion at an activity often considerably lower than the initial activity. In addition to the need for stable activity, equally important is the requirement of structural stability. Most catalysts are charged by pouring into a reactor from a significant height, incurring breakage of the weakest particles. Once in service, the catalysts are often then subjected to a significant pressure drop through the bed, leading to further breakage with a resulting increase in pressure drop, and so on. Other catalysts, either by accident or design, may contact liquids and need to be able to withstand this happening without degrading to a sludge. Examples include boiler failure with shift catalysts and any 'trickle-bed' liquid phase application as employed in some hydrogenations. Catalysts used in fluid-bed converters need to be sufficiently strong to withstand excessive attrition in use, but must not be so hard as to cause erosion of the converter.

Catalysts come in many shapes and sizes and consist of many different materials. Some may consist of, or be derived from, naturally occurring materials such as zeolites and clays, whereas others may be prepared via synthetic microcrystalline minerals such as malachite or pyroaurite. Others again, particularly though not exclusively derived from platinum group metals, are prepared via aqueous impregnation of a support which, while often supposed to be inert, in reality contributes significantly to both the catalytic activity and selectivity of the product. Thus, a fully developed catalyst is an 'effect' chemical and its true value is far in excess of the value of its component parts. Most catalysts are used in pelleted form, either as tablets or rings, but other forming methods such as extrusion or granulation also find extensive application. Powdered catalysts, particularly in the form of microspheres to minimize attrition losses, are used in fluid bed processes such as the oxychlorination of ethylene where heat transfer can prove to be a problem.

The subject of catalysts is so vast that it is completely beyond the scope of a single volume and no attempt has been made to offer comprehensive coverage. Large areas such as polymerization, oxidation and enzyme catalysis have been omitted, and instead a small number of topics has been dealt with in some detail. The topics themselves are widely diverse to give the new reader a broad introduction to catalysis. The first and second chapters are quite general in content, applying to all forms of heterogeneous catalysis. The remaining chapters are more specific and discuss topics of current general interest.

In Chapter 1, Mr Chinchen describes some of the problems encountered in the activitation and more specifically the deactivation of catalysts, the re-activation of some deactivated catalysts, active-site blocking, chemical sintering and thermal sintering.

The last two decades have seen the introduction of a vast array of physical techniques for the study of solids which are invaluable to the catalyst chemist and which complement the more traditional methods such as kinetic studies and micromeritic measurements. Dr Foord describes a selection of these new techniques in Chapter 2, with particular emphasis on extended X-ray absorption fine structure, solid state nuclear magnetic resonance spectroscopy and microscopy. The chapter is completed by a description of the surface scientists' approach to catalysis using well defined crystal structures under ultra high vacuum conditions.

Although zeolites have only found application in catalysis since 1962 with the introduction of the Mobil cracking catalysts, they have become the subject of enormous research effort all around the world, particularly since the introduction of the high silica products of the ZSM-5 type. In Chapter 3 Dr Spencer traces the history, structure, development and use of zeolites with many specific and current examples.

Chapter 4, by Dr Twigg and myself, is rather different: the catalytic aspects of two mature processes are discussed, combining developments leading to the current processes for ammonia and methanol production with descriptions of some of the catalysts used at the various stages of the process.

All the contributions up to this stage in the book deal with the subject of heterogeneous catalysis. It must not be forgotten that not only is homogeneous catalysis extremely important in organic synthesis, but several large tonnage chemicals are manufactured via homogeneous catalysis. The problems associated with product and catalyst recovery are more than compensated for by the high selectivity which the homogeneous process endows. It is this advantage in selectivity which is likely to be increasingly exploited in the future, particularly in the synthesis of small tonnage speciality chemicals requiring assymmetric or stereoselective synthesis. In the final chapter of this volume, Dr Whyman reviews those applications of homogeneous catalysis which have achieved full production status, including examples of both large and small tonnage products.

Catalyst chemistry has long been seen as a 'black art'. Indeed, many of the discoveries and their subsequent development were achieved by empirical means rather than through a detailed understanding of the chemical processes taking place. This, I believe, is changing and the wealth of knowledge being generated about catalysts and catalytic processes should allow research into new catalysts to proceed in a more positive, and hopefully fruitful, manner.

In conclusion, I wish to thank the authors for their individual contributions, my colleagues and friends who have read the manuscripts, the typists who produced them, and everyone else involved in the production of this volume.

J.R. Jennings

1 Activation and deactivation of heterogeneous catalysts

G.C. Chinchen

1 Activation of catalysts

The term 'activation of catalysts' usually refers to any specific process or treatment which is carried out after charging the catalyst to the reactor and before bringing the reactor on-stream. Activation is really the final stage in catalyst manufacture, and it is carried out in the reactor rather than the catalyst factory for convenience of handling and transporting the catalyst. It is usually necessary to convert the catalyst from its air-stable precursor form into its active state; although stable under normal reaction conditions the active state is often not air-stable. Unlike catalyst deactivation, which usually occurs in spite of all the designs and controls of the catalyst maker and user, catalyst activation should be a well controlled, reproducible process carried out under carefully designed conditions.

1.1   Reduction procedures

One of the commoner methods of activation is the reduction of metal oxides/hydroxides/carbonates to the metal. The reduction is usually carried out relatively slowly under carefully controlled conditions; this limits any temperature rise and minimizes exposure of the catalyst to water vapour produced in the reduction. Both high temperatures and high levels of steam can cause premature sintering of the catalyst, and the catalyst manufacturer usually provides a reduction procedure for his catalysts with this in mind.

The catalyst must have physical properties which can withstand any damage during the activation process so that the activated catalyst retains adequate strength. A typical procedure for the activation of nickel (oxide/cabonate) catalysts is to purge the catalyst bed of air with nitrogen and then warm up to about 400°C whilst passing nitrogen to decompose any carbonates present. After cooling to about 330°C, hydrogen is bled into the nitrogen and a hot spot will then develop which passes through the bed. The hot spot is kept well below 400°C by controlling the H_2 concentration until the catalyst is sufficiently stable that pure hydrogen can be admitted without the temperature exceeding 400°C.

Basically similar procedures—with suitable adjustments to temperatures and reduction gas compositions—may be used to reduce copper, cobalt and iron catalysts. In the latter case, the active species required is sometimes Fe_3O_4, and the activation then has to be carried out with a defined level of steam present (e.g. water gas shift catalysts).

In some cases, particularly with nickel catalysts, the temperatures required to effect reduction/activation are higher than can be achieved in the reactor, and pre-reduced catalysts may then be used. Here the catalyst is reduced by the manufacturer and is then either stabilized (surface oxidized under mild conditions) to make it safe to handle in air, or is supplied in a protective liquid or solid (usually one related to the process for which the catalyst will be used).

1.2   Other methods of activation

An example of a somewhat unusual activation procedure is that applicable to Raney metal catalysts. These catalysts begin as alloys, usually of the catalytic metal and aluminium although other promoting metals may also be present, e.g. zinc, chromium. They are activated by leaching with sodium hydroxide solution and are used immediately after washing free of alkali with water. This activation with alkali dissolves away a large proportion of the aluminium leaving the catalytic metal in high area. The properties and activity of the Raney catalyst depend to some extent on the method of activation. Raney nickel is probably the most commonly used nickel catalyst and its properties and uses have been reviewed.[1] The nickel crystallite size is typically 4–8 nm and the catalyst still contains some aluminium. Freshly prepared it also contains hydrogen, most of which can be removed by heating. Raney cobalt, iron and copper have been prepared by similar techniques and recently Raney copper–zinc catalysts have been prepared and shown to be active for methanol synthesis.[2] The residual zinc in these catalysts was found to act as a promoter.

The production of high area metal catalysts from alloys can be achieved in ways other than the Raney method of leaching the non-catalytic component from the alloy. Thus, copper catalysts active for methanol synthesis have been made from alloys of copper and thorium essentially by the oxidation of the thorium to thoria.[3] As in the case of Raney catalysts this technique has been applied to a number of metals—mostly Group 7 metals in combination with yttrium, rare earth metals and actinides—and the activation of these alloys carried out in various gas mixtures.[4]

Catalysts for some reactions require activation pre-treatments specific to the reaction conditions. For instance, nickel molybdenum and nickel tungsten catalysts used in hydrofining processes to remove sulphur from hydrocarbons frequently require pre-sulphiding before operation. Like reduction, the sulphiding reaction is exothermic and similar care in carrying out the operation is necessary. Carbon disulphide or H_2S are frequently used as the pre-sulphiding agents. Similarly, catalysts used in halogenation or oxyhalogenation reactions may require a halogenation stage in their activation. Reaction selectivity can sometimes be controlled by the partial posioning of a catalyst, in which case the activation procedure will include some means to achieve the partial poisoning.

1.3   Reactivation of catalysts

Following the successful activation of a catalyst, it is likely to suffer from one or more deactivation processes (described below) during its operating life. When its performance has fallen below acceptable levels, the catalyst either has to be replaced or, in some cases, it may be possible to reactivate it (more commonly called regeneration) for a further period of useful life. Whether a catalyst can be reactivated or not depends on the reason for its loss of activity.

Sintering is generally an irreversible process: it is possible to find ways of redispersing a sintered catalytic species, but the conditions required to achieve this usually make it unfeasible. Because sintering is irreversible, processes for reactivating catalysts which have been poisoned or fouled are also limited to conditions which do not cause rapid catalyst sintering.

Poisoning may be reversible, in which case the removal of the poison from the feed stream should result in steady reactivation of the catalyst. In some cases this reactivation can be accelerated by temporarily operating at higher temperatures or under different feed compositions. If other problems have been caused by the poisoning (e.g. fouling by carbon deposition due to reduced activity of poisoned catalyst in steam reforming processes) then further reactivation procedures may have to be undertaken or reactivation may prove impossible. Severe poisoning may only be reversed by adopting more specialized conditions to remove the poison, a procedure which always puts the catalyst at risk from sintering.

Fouling is the deactivation process which is most likely to afford successful reactivation procedures. The deposition of 'coke' on hydrocarbon processing catalysts is very common. It can often be removed by taking the catalyst off stream and heating it in an atmosphere with a controlled oxygen level; the coke is then burnt off as gaseous CO and CO_2. Again, the large exotherm which results has to be controlled—usually by using very low concentrations of oxygen—to avoid sintering of the catalyst. The regeneration of coked catalysts in this manner is probably the most common and most successful catalyst reactivation process in use, and the large literature on all aspects of this process has been well reviewed and described.[5]

Catalysts fouled by other materials than coke have also been successfully regenerated. Plant mishaps often result in catalyst beds being subjected to fouling by a wide range of impurities, e.g. in the manufacture of ammonia, potash may be deposited on water–gas shift catalyst from upstream steam reforming catalyst. In many cases the catalyst has been retrieved by taking it off stream and washing it in a suitable solvent for the impurity which fouled the bed. When taking catalysts off line to carry out such procedures it is often necessary to passivate them first, i.e. careful oxidation to allow damage-free contact with the air. An unusual but successful regeneration of a fouled/poisoned catalyst is acid washing to dissolve calcium impurities which had migrated to the surface of a cobalt oxide ammonia oxidation catalyst.[6]

2   Deactivation of catalysts

Most catalysts used in heterogeneous catalytic processes lose activity during operation, and the time taken for their activity to fall to an unacceptable level ('catalyst life') varies with the process and the conditions employed. The design of catalysts with a guaranteed operating life, whilst a very desirable objective, is also very difficult to realize. Obtaining reliable information on catalyst life is probably the most difficult facet of catalyst development because life testing makes heavy demands on resources and may prove unreliable. Usually some kind of accelerated ageing technique is used—depending on the dominant deactivation mechanism, there are many ways of designing accelerated ageing. Care is necessary in the design and interpretation, however, as a change in deactivation mechanism at these accelerated conditions can present totally misleading conclusions.

2.1   Type of deactivation process

The practical operating life of a catalyst is governed generally by three types of deactivation process: chemical, thermal and mechanical.[7]

Chemical deactivation processes involve a chemically-induced change in the catalyst with resulting change in catalyst activity. This change may be the adsorption, or deposition on, or the reaction of what is generally termed a poison (*poisoning*) with the catalyst surface. The poison may be an impurity or in the case of self-poisoning, a reactant, intermediate or product. A common example of self-poisoning is catalyst coking (*fouling*) in which carbonaceous material formed as a side reaction is deposited on the catalyst and prevents access of reactants to the surface. Such loss of activity may also be caused by material physically blocking the catalysts' pore structure (*pore-plugging*). Either impurities or reactants/products may, by chemical reaction with the catalyst, induce a restructuring of the surface (*poison-induced sintering*).

Thermal deactivation processes include loss of active area generally by crystal growth (*sintering*) and a host of other phenomena of a more specific nature such as alloying, segregation, volatilization and metal-support interactions. It is often difficult to separate entirely, thermal processes from chemical ones as many thermal processes are strongly affected by the chemical environment of the catalyst, as in poison-induced sintering.

Mechanical deactivation processes include loss of strength and physical breakdown of catalysts, causing excessive pressure drop in packed beds and attrition leading to excessive fines in fluid bed catalysts.

In the space available it is not possible to provide an exhaustive description of the phenomenon of catalyst deactivation, nor a comprehensive discussion of the literature on this subject. Instead, an attempt will be made to introduce the major topics in a systematic way and develop them by quoting appropriate examples, leaving further detail to some of the excellent reviews of these topics which have appeared in the last 30 years or so.[5,7-14] The phenomenon of catalyst deactivation can be systematically discussed under the three major headings of sintering, poisoning and fouling.

3   Sintering

For heterogeneous catalysis the term 'sintering' has a more extensive meaning than in general ceramics usage; here it includes all those temperature-dependent processes which result in growth of the catalyst particles and in loss of active surface area. In the case of a supported catalyst such processes may lead to an overall loss of area of the catalyst support material or may merely result in a loss of dispersion of the active metal crystallites. Experimentally determined activation energies for sintering are generally high, implying a rapid increase in sintering rate with increase in temperature.

3.1   General description

For a non-defective, stoichiometric single phase in an inert atmosphere the observed sequence of events in sintering as the temperature is increased is as follows.

(a)   Surface smoothing: near the Huttig temperature (0.2–0.3 T melting °K) the onset of surface diffusion leads to the disappearance of high index planes.

(b)   Necking: the formation of concave bridges between particles in contact leads to the appearance of aggregates.

(c)   Aggregate consolidation: particles joined by necks transform into larger single particles with further smoothing and some recrystallization near the Tamman temperature (0.5 T melting °K).

(d)   Metastable state: the particles are large and of much the same size so further movement is minimal. If the phase is extended on another unsinterable support, growth ceases because the particles are well separated and diffusion on the surface of the support over large distances is slow.

(e)   Various phenomena occur, e.g. surface defects, vaporization; near the melting point volume defects occur.

The first four stages have been observed in the growth of small gold crystals on inert supports.[15]

3.2   Sintering mechanisms: general

Several detailed mechanisms have been proposed for sintering.

(a)   Evaporation–condensation: the higher vapour pressure over a convex surface will lead to a tendency for material to evaporate and condense at concave intersections between particles.

(b)   Volume diffusion: this may occur by a variety of possible mechanisms, e.g. migration of interstitials (Frenkel defects), migration of vacancies (Schottky defects).

(c) Surface diffusion: a mechanism involving the migration of atoms across the particle surface.

(d)   Grain boundary diffusion.

Mechanisms and kinetics for sintering processes have mostly been derived from studies of relatively large particles (0.1–100 μm). Examples are found in the work of Coble.[16] At its lower extremity this size range overlaps that of the larger crystallites found in catalysts, e.g. nickel in steam reforming catalysts. Little has been done with very small particles ($<$ 0.1 μm), but surface diffusion is known to be the principal mechanism controlling the aggregation of small two-dimensional crystals of gold on MoS_2,[15] the more so the smaller the crystals.

Studies of the rates of sintering have traditionally been correlated by power law functions of the form

$$\frac{dD}{dt} = \frac{k_1}{D^{(m-1)}} \tag{1}$$

or in integrated form

$$D^m - D_0^m = k_2 t \tag{2}$$

where D_0 is the initial particle size, D is the particle size at time t and k_1, k_2 and m are constants. Alternatively, if the surface area is used to measure the degree of sintering then,

$$-\frac{dA}{dt} = k_3 A^n \tag{3}$$

or in integrated form $(n \neq 1)$

$$\frac{1}{A^{n-1}} - \frac{1}{A_0^{n-1}} = k_4 t \tag{4}$$

where A_0 is the initial area, A is the area at time t, and k_3, k_4 and n are constants. For spherical particles the exponents in equations (1) and (3) are related by

$$m = n-1 \tag{5}$$

The values of the exponents obtained in sintering studies have been used as a guide to the specific sintering mechanisms involved. For instance, in studies of the rates of sintering of relatively large particles of single phases the predominant mechanisms have been ascribed typically to evaporation–condensation for $m = 3$, volume diffusion for $m = 5$ and surface diffusion for $m = 7$. Such principles are not strictly applicable to the situation in most catalysts, because catalysts are not usually single phases, and because the initial sizes of catalyst particles are often so small that a large proportion of their energy is surface energy, a term which can be virtually neglected in large-particle analyses.

With these reservations, studies of the sintering of water gas shift catalysts (Fe_2O_3/Cr_2O_3) have given values of the exponent $n = 7.2$ and 8.3[17] for two catalysts with differing Cr_2O_3 contents, with an activation energy of 272 kJ mol.$^{-1}$ A more recent study[18] found values of $n = 3$ for the initial activity decay and $n = 5.7$ for the later stages, with an activation energy of about 200 kJ mol^{-1}. A speculative model for the two stages of decay was proposed, in which the catalyst after reduction consisted of some crystallites of Fe_3O_4 in direct contact or nearly so and others in positions where contact was prevented by Cr_2O_3 (or Fe/Cr_2O_4)

crystallites. The former sinter rapidly, whereupon further growth requires transport of Fe_3O_4 which is likely to be a slower process.

More effort has recently been devoted to the problem of agglomeration or
decreased dispersion of metal crystallites in supported metal catalysts. The
purpose of the support in such systems is to act as a spacer and reduce the
growth/movement of the metal crystallites. Inevitably if the support itself sinters,
then the spacer action fails. Williams[19] has provided quantitative results for the
sintering of both metal and support for a Ni/Al_2O_3 catalyst. Both metal and total
catalyst area show a rapid initial fall during the early stages of sintering, whilst
there is only a much slower fall during the later stages. The magnitude of the rapid
initial fall increases markedly with increasing temperature.

Table 1.1. Values of sintering exponent n for supported platinum catalysts

Author	Ref.	Value of n	Conditions
Maat & Moscou	23	2	Air/780°C 0.6% Pt/γ Al_2O_3
Hermann *et al.*	20	2	N_2/564–625°C 0.375% Pt/γ Al_2O_3
Gruber	22	6	H_2/500°C 1.1% Pt/η Al_2O_3
Hughes *et al.*	21	8	H_2/483–538°C 0.4% Pt/η Al_2O_3

Most of the work quoted in the literature for the sintering of supported metals
is concerned with platinum catalysts[20-23] for which a general review is given by
Flynn & Wanke.[24] Values of the exponent n for the above studies are given in
Table 1.1. A more exhaustive tabulation of available literature values both for the
exponent n and for the apparent activation energies for sintering is given by
Hughes.[5] The values of n vary between 2 and 16 and the activation energies whilst
all generally high, vary from some $50\ kJ\ mol^{-1}$ to in excess of $350\ kJ\ mol^{-1}$. The
most important factors which affect the rate of sintering are (i) temperature and
(ii) the atmosphere under which sintering occurs. The rate of sintering is larger in
oxygen-containing atmospheres.

3.3   Sintering mechanisms for supported metal catalysts
Two distinct mechanisms for the growth of metal crystallites on supports have
been proposed. A model based on particle migration and coalescence was
published by Ruckenstein & Pulvermacher,[25,26] whilst a model based on the
transfer of metal atoms individually from one particle to another (interparticle

transport) was proposed by Flynn & Wanke.[27,28] There has been much discussion about these mechanisms.[29-31]

3.3.1   Particle migration. This model is based on the reasoning that above the Tamman temperature the metal crystallites have an increased mobility. They migrate as crystallites over the surface of the support because the interactions within the crystallites between atoms are stronger than those between the crystallites and the support. Two different situations may arise: (i) 'diffusion control', when two particles coming into contact merge to form a single unit, the time taken being short compared with the migration time; (ii) 'sintering control', when the merging process is slow compared to the diffusion process. The initial analysis by Ruckenstein & Pulvermacher[25,26] gave values of the exponent n of 2–3 for sintering control and of 4–6 for diffusion control. Flynn & Wanke[24] suggest that whilst some particle migration may occur in the initial stages of sintering, this is unlikely to be the dominant mechanism for larger particles and they devised a model based on an alternative mechanism—atomic migration.

3.3.2   Atomic migration. This mechanism is seen to occur in three steps: (i) movement of metal atoms from the crystallite to the surface of the support; (ii) migration of metal atoms on the support surface, and (iii) capture of migrating metal atoms by collision with a metal crystallite.

3.3.3   The predominant mechanism. As already mentioned, there has been much discussion about which of these two migration mechanisms is predominant in catalysts. Arguments against particle migration include that of Flynn & Wanke,[27] that metal crystallites sinter to sizes greater than those of the support particles, which is difficult to explain by particle migration. It has also been argued that observed values for the exponent n of up to 15 cannot be accounted for by the particle migration model which, as originally presented, gave the exponent n an upper limit of 8.[26] Ruckenstein & Dadyburjor[32] have pointed out that this limit can be revised upward, if it is assumed that the diffusivity is inversely proportional to higher powers of the particle size. A major problem in the atomic migration model is the initial energy required to remove the atom from the crystallite to the surface of the support. The very high heat of sublimation of platinum and the small heats of adsorption of platinum atoms on supports, suggest that growth by atomic migration would be extremely slow. An oxygen-containing atmosphere allowing some surface oxidation would improve this situation, as PtO_2 has a significant vapour pressure at higher temperatures. The energy required to remove a Pt atom to the gas phase (as a PtO_2 molecule) would then be considerably less than the sublimation heat of Pt. Electron microscopy has provided evidence for crystallite migration.[33,34] 'Islands' of copper, silver and gold have been observed to move and coalesce on supports such as amorphous carbon at temperatures as low as 500 K. Supported metal crystallites are frequently in the same size range as these 'islands', i.e. 1–10 nm.

Granqvist & Burhman[31] have argued that it is possible to distinguish between the two theories in a particular case by examining the particle size distribution. Their analysis of all previous work[35] on both interparticle transport and crystallite migration concluded that the two theories predict distinctly different distributions. The crystallite migration model predicts removal of very small crystallites with a distribution 'tailing' on the high diameter side of the peak, whereas the interparticle transport model predicts a substantial tail on the low-diameter side. They maintained that the accumulated evidence pointed almost unequivocally in favour of coalescence growth rather than single atom migration.[36] Wanke, however, argues that some initial distributions give log-normal functions on sintering via interparticle transport.[30] Wynblatt & Gjostein[29] consider that both mechanisms contribute, migration dominating for small crystallites while interparticle transport occurs for the larger ones. Richardson & Crump[37] consider that the crystallite size distribution history may be the only reasonable method for differentiating the two models. Most data have come from electron microscopy and are uncertain due to sampling and statistical errors, but some studies do point to a particular mechanism. For instance, Nakamura[38] observed the disappearance of small platinum crystallites on charcoal and Bett *et al.*[39] used size distributions to support crystallite migration. Richardson & Crump[37] applied the method of magnetic granulometry[40] to the sintering of nickel on silica and observed that at 723 K sintering occurred with the disappearance of small crystallites, the distribution reaching a limiting log-normal shape independent of initial distribution. They interpreted these results as particle migration and found an activation energy of 200 kJ mol^{-1}, so much smaller than the energy for atom detachment (431 kJ mol^{-1}) that interparticle transport seemed unlikely. The exponent *n* had a value of 10 within the predicted range for particle migration if adjustments were made for facetting. At higher temperatures of 873 K bimodal distributions resulted, suggesting an influence of pore size distribution. Crystallites with dimensions close to those of the pore may be stabilized against sintering. Hughes[5] summarizes the current position on sintering as having the following pattern. For very small particles (< 20 nm) growth occurs predominantly by particle migration, with particle diffusion rate controlling. For larger particles, growth occurs by atom migration on the surface. Kuo *et al.*[41] examined the sintering of silica-supported nickel catalysts, and the changes in particle size distribution during sintering below 700°C suggested a particle migration mechanism, while at 800°C atomic migration seemed to predominate.

3.4 General factors which influence sintering rates

For metals, the predominant sintering mechanism in the bulk is vacancy diffusion which suggests a relationship with cohesive energy (and thus also with the melting point). Hughes[5] gives the following increasing order of stability for metals:

Ag < Cu < Au < Pd < Fe < Ni < Co < Pt < Rh < Ru < Ir < Os < Re

For oxides, sulphides and halides there is generally a relationship between lattice energy and melting point so the melting point is again a reasonable index of sinterability.

Modifications of the stoichiometry of solid phases by heat, atmosphere, or solid state reactions can have marked effects on the rate of sintering. For example, zinc oxide loses oxygen on heating in air and this non-stoichiometry is accommodated by the formation of interstitial zinc atoms or oxygen vacancies which assist sintering. The process is accelerated in a reducing atmosphere and retarded in an oxidizing one. In general, compounds in which the metal ions are in their highest valency become oxygen deficient on heating and sinter faster in reducing atmospheres, while compounds with metal ions in their lowest valencies become metal deficient and sinter faster in oxidizing atmospheres.

The adsorption of polar molecules such as H_2O, H_2S, HCl, NH_3 can increase the mobility of materials in many ways. The sintering of copper metal in a water gas shift catalyst[42] is greatly accelerated by extremely low levels of HCl, via the formation of highly mobile copper chloride. The increase in rate of surface diffusion of copper in the presence of traces of chlorine has been measured.[43] A similar effect of arsine on platinum has been reported.[44]

In stabilized catalysts where a refractory material spaces the crystallites of the active phase in order to inhibit sintering, the intimacy of mixing is an important factor. In a stabilized catalyst of this type, the stabilizer can only fulfil its role if there are sufficient crystallites of stabilizer relative to those of sinterable material. Andrew[45] has reported that if distribution is good, experimental data provide the following rough rule:

$$D_{sint}/D_{stab} = 7 \, V_{sint}/V_{stab} \tag{6}$$

where D_{sint} is the smallest crystallite size of sinterable material when sintering is complete in a catalyst containing relative volumes V_{sint} and V_{stab} of the two species, the size of the stabilizer being D_{stab}. This relationship is shown in Fig. 1.1, which also shows catalyst activity assuming this is directly proportional to the surface area of active species, i.e. activity $\alpha \, V_{sint}/D_{sint}$. The relevance of this rule to the design of a copper-based water gas shift catalyst has been discussed.[46] Typically, such a catalyst may contain 20% copper by volume of initial size c. 5 nm, together with refractory components of initial size also c. 5 nm. Provided there is no growth of the refractory the rule implies that the copper will be restrained to a size of < 10

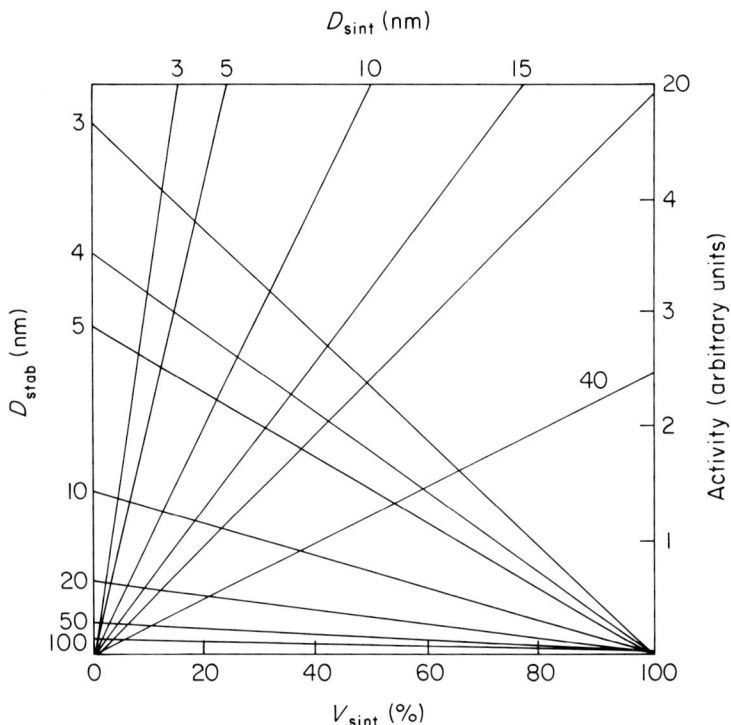

Fig. 1.1. D_{stab} is represented by the lines intercepting the left hand axis, the appropriate line being traced to the value of V_{sint} used. The point of interception gives D_{sint} by reference to the lines intercepting the top axis, whilst the catalyst relative activity is read from the right hand axis.

nm, giving a stable activity of 2.3 units. If the volume per cent of copper were increased to say 40% in an attempt to increase the activity, the copper would then sinter to a size of c. 20 nm, giving a stable activity of only 1.7 units. To stabilize this higher copper content would mean reducing the refractory crystallite size to < 3 nm, probably an impossible task.

4   Poisoning

Poisoning is a form of catalyst deactivation caused by adsorption, or deposition on, or reaction of small amounts of material with the catalyst surface, usually specific to a particular catalyst, generally termed a poison. The poison may be a contaminant in the feed but it may also be a reactant intermediate or product of the desired or an undesired reaction. Many poisoning processes are irreversible but there are some which are reversible, e.g. poisoning of ammonia synthesis catalyst by water vapour (Fig. 1.2), and the poisoning of nickel catalysts by water vapour.[47] Ultimately, the only way to prevent impurity poisoning is to remove the

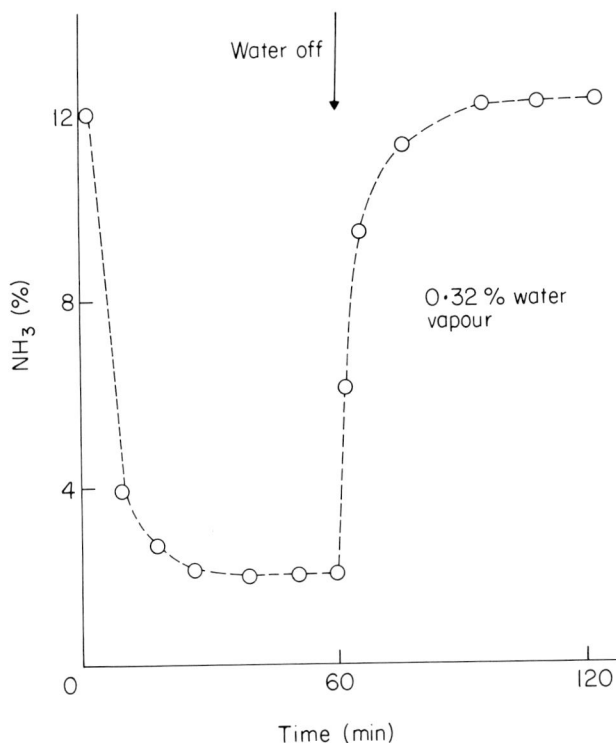

Fig. 1.2. Reversible poisoning of ammonia synthesis catalyst.

impurities responsible in the feedstock, or use a guard bed of catalyst; on the other hand, self-poisoning may be impossible to eradicate.

In practice, some degree of poisoning is generally tolerated. The front between already-poisoned catalyst and unpoisoned catalyst—where reaction starts—travels down the catalyst bed at a known steady rate, so adequate catalyst life can be ensured by having sufficient catalyst bed length.

4.1 Poisoning of metallic catalysts

Maxted[48] in a review, identified the metals susceptible to poisoning and the primary materials responsible for such poisoning. The metals most susceptible to poisoning are Fe, Co, Ni, Ru, Rh, Pd, Ir, Pt and Cu, and the common poisons for these metals were identified in two classes.

(a) Either the free elements of Groups 5B and 6B or molecules containing these elements, viz N, P, As, Sb, O, S, Se, Te.

(b) Molecules containing multiple bonds such as carbon monoxide or strong adsorbates such as benzene.

The degree of toxicity is related directly to the electronic state of the toxic element in the molecule. If the normal valency orbitals are saturated by stable bonding to other elements in the molecule, then there is no toxic activity; whilst if there are unshared electron pairs or empty valency orbitals, then chemisorptive bonding to the metal is possible and poisoning occurs. Under severe reducing conditions some non-toxic structures may be converted to toxic forms, particularly with As and Sb. Thus, a substance may poison a catalyst for one reaction but not another. For instance, arsenic compounds generally poison platinum catalysts used for hydrogenation reactions because the arsenic is present as arsine; however, arsenic compounds are not toxic for platinum catalysts used for the decomposition of H_2O_2 because the arsenic is then present as arsenate due to the strong oxidizing conditions.

Poisons which are toxic due to the presence of multiple bonds are of interest since they form a class of compounds which are readily hydrogenated over transition metal catalysts. A wide range of poisoning efficiencies occur for different unsaturated compounds; if two such substances are present, the 'poisoning' effect of one on the hydrogenation of the other may vary from equal competition for active sites to complete suppression of one reaction by even small amounts of the other. Such systems offer much scope for selective hydrogenation, e.g. acetylene to ethylene in the presence of carbon monoxide.

4.2   Poisoning of non-metallic catalysts

The majority of studies of the poisoning of non-metallic catalysts involve cracking catalysts consisting generally of acidic oxides which are poisoned by a variety of organic bases. A study by Mills et al.[49] showed how poisoning could be used to investigate the origins of catalyst activity. Silica–alumina catalysts were poisoned by a variety of organic bases and the activity measured by cumene cracking. The organic bases were highly selective poisons, activity falling exponentially with poison concentration. For example, 4% coverage by quinoline decreased the activity by a factor of seven. Mills concluded that by far the major part of the surface made no contribution to the activity of the catalyst. The acidic nature of silica–alumina catalysts was revealed by the adsorption and desorption of quinoline and, whereas quinoline is completely desorbed from pure silica on heating to 315°C, on silica–alumina a portion of the quinoline is irreversibly adsorbed. The cracking activity is related to the amount of quinoline which can be irreversibly adsorbed and, hence, to the acidity of the surface caused by the interaction of the silica and alumina. The toxicity of the poisons was related to their basicity, except for piperidine which is less toxic than expected because it is cracked during the reaction.

Non-metallic catalysts for ammonia oxidation to nitric oxide, typically cobalt oxide catalysts, are poisoned by sulphur dioxide. A radiochemical study by

Apelbaum[50] showed that this poisoning was reversible, that the reaction on the poisoned catalyst took place in the external diffusion range so that the poisoning was only shown by a change in apparent selectivity of the catalyst (NO relative to N_2) and that a fall in selectivity was observed for sulphur concentrations as low as 40 μg m^{-3}. This corresponded to a coverage of the catalyst by sulphur of about 10%. These facets of the poisoning of cobalt oxide catalysts were confirmed[51] and the threshold concentrations of SO_2 at which an observable fall in catalyst selectivity to NO occurred were measured as a function of temperature. Whilst the catalyst would tolerate some 1500 μg m^{-3} at 800°C, the threshold level was down to about 200 μg m^{-3} at 750°C and only about 50 μg m^{-3} at 700°C. By way of comparison, platinum/rhodium gauze catalysts for this reaction were not affected by SO_2 at 750°C up to levels of 3000 μg m^{-3},[51] which is one of the reasons why they are preferred to oxide catalysts for this reaction.

Table 1.2. Common poisons for some industrial catalytic processes

Process	Catalyst	Poison	Comment
Steam reforming	Supported nickel (typically Al_2O_3 as support)	Sulphur	Generally reversible:[52] threshold value lower at lower temperatures.[52,53]
		Arsenic	Practically permanent.
		Halogens	Similar to S, generally reversible.
Shift catalysts	Supported copper e.g. $Cu/ZnO/Al_2O_3$	Halogens	Probably poison-induced sintering—irreversible. Threshold levels—fractions of ppm. Poisoning pore diffusion limited.
		Sulphur	Reversible poison but generally requires much higher levels than halogen to produce same deactivation.
Ammonia synthesis	Supported promoted iron e.g. $Fe/K_2O/Al_2O_3$	Sulphur	Permanent poisoning by low levels of S, sensitivity to S reduced by having CaO present in formulation.
		Oxygen compounds e.g. H_2O, CO	Temporary poisoning.
Ammonia oxidation	Cobalt oxide	Sulphur dioxide	Reversible poisoning evidenced by fall in catalyst selectivity to NO.
	Pt/Rh gauge	,,	Reversible poisoning but requires very high levels of SO_2 to cause fall in selectivity.
SO_2 oxidation	Supported platinum	Any metallic compounds, halogens, phosphorus, arsenic, mercury	Very susceptible to poisoning. Halogen poisoning sometimes reversed by heating the catalyst in air.
	Vanadium catalysts	,,	Very much less suceptible to poisoning than Pt. Many orders of magnitude greater levels of poison can generally be tolerated.

By way of summary, common poisons for some particular catalytic processes using both metallic and non-metallic catalysts are given in Table 1.2.

4.3   Quantitative aspects of poisoning

Experiment has shown a large number of quantitative relationships between the catalyst activity in a system and the amount of poison present. Poisoning inter-actions have been labelled as 'selective' or 'non-selective'; when the first amount of poison causes a greater fall in the activity than later amounts the interaction is described as 'selective'. This may indicate that the poison is adsorbed first on those parts of the surface which constitute the most active sites, but is often a con-sequence of mass transfer limitations which will be discussed later.

An example of non-selective behaviour is the arsine poisoning of cyclo-propane hydrogenation over Pt.[54] Here a good linear relationship between activity and level of poison was observed expressible as:

$$a = a_0 - \beta C_p \hspace{4cm} \text{Linear} \hspace{2cm} (7)$$

where a_0 is activity, a is initial activity, C_p is poison concentration and β is poison coefficient.

By contrast, extremely selective behaviour is shown in the carbon monoxide poisoning of ethylene hydrogenation over Cu.[55] The activity is strongly affected by only small amounts of poison and the effect tapers off at higher poison concentration. The relationship in this case is exponential, but such selective or non-linear behaviour has been represented empirically by a variety of expressions such as:

$$a = a_0 \exp(-\beta C_p) \hspace{3cm} \text{Exponential} \hspace{1.5cm} (8)$$

$$\frac{1}{a} = \frac{1}{a_0} + \beta C_p \hspace{3.5cm} \text{Hyperbolic} \hspace{1.5cm} (9)$$

There is no particular fundamental significance associated with any of these forms, but they may be used to infer deactivation rate equations and hence indicate something about the poison mechanism. For instance, in the thiophene poisoning of a platinum catalyst for the liquid-phase hydrogenation of crotonic acid,[48] the behaviour is linear down to a relatively low activity, whereafter the curve becomes non-linear and more exponential in character. This behaviour could be explained if more than one site was required for adsorption of the thiophene.

Levenspiel[56] recognized that a general kinetic equation for quantitatively describing the behaviour of a catalyst subject to poisoning, or indeed to any form of deactivation, would have to be separable in form to be useful. Separable in this case means that the reaction rate function can be expressed as a product of two terms: a kinetic term which is time-independent and an activity which is time-dependent. The separable form allows the expression of an instantaneous rate of reaction as the product of independent individual factors. Fortunately, many real catalytic deactivation mechanisms may be considered as additional chemical processes to the main catalysis, and hence lead to separable rate equations which do correspond to physical or chemical reality, even when an apparently physical process such as sintering is concerned.

Szepe & Levenspiel[56] proposed a general power law form for deactivation kinetics, and Levenspiel and his students[57] have gone on to show that, by varying the value of the deactivation order, these kinetics can cover most types of observed catalyst decay—even to complex situations in which mass transfer limitations are interacting with the catalyst deactivation. Separable deactivation kinetics have been used almost exclusively to describe catalyst deactivation: only in the cases of non-ideal surfaces[58] and non-linear site balances[59] have separable kinetics proved inappropriate.

5   Fouling

Fouling is a process in which, in general, much larger amounts of material cause deactivation than is the case in poisoning. Typical of fouling is coke formation on catalysts commonly found in catalytic reactions involving hydrocarbons. Highly unsaturated species of high molecular weight are adsorbed onto the catalyst, polynuclear aromatics being especially potent, and after adsorption further condensation reaction leads to the formation of a hydrogen-deficient 'coke'. The coke deposit in this case originates from reactions occurring over the catalyst and is not an impurity, so that fouling by coke, unlike poisoning, cannot be eliminated by purifying the feed or using a guard bed of catalyst. Coke formation can be reduced to a minimum by careful selection of operating conditions and optimization of the catalyst properties. Regeneration is often possible by oxidation of the coke in steam or diluted air but may cause some loss of activity due to sintering of the catalyst. In other cases undesired polymerization reactions can lead to the formation of gums and waxes and, in another class of fouling reactions, metal sulphides are deposited arising from reaction between organometallics in the feed and sulphur-containing molecules during hydrotreating processes.[60]

In all cases of fouling, active catalytic area is lost without change of crystal size of the active component; the area is lost essentially due to encapsulation by a solid which lacks the required catalytic properties. Fouling occurs either as a result of undesired surface reactions yielding materials which foul the surface or by the introduction of materials in the reaction feed which lead to something being physically deposited on the surface. If the quantities of deposit become excessive, not only can the active surface be covered but the diffusional characteristics of the porous catalyst pellets can be impaired and in the limit the void space between pellets may be blocked, effectively shutting down the reactor.

5.1   Fouling by coke formation

A number of studies have been made of the nature of the coke deposit on catalysts. For oxide catalysts, including cracking catalysts, it was established by microscopy[61] that the coke was in the form of filmy aggregates of less than 10 nm particle size. It is generally agreed that the coke deposits are hydrocarbon in composition but with only a small amount of hydrogen ($C_1H_{0.4}$–C_1H_1). In general, unless large amounts of coke are formed the physical properties of the catalyst are not changed. More recent work using electron microscopy has suggested[62,63] that coke deposition on metal and supported catalysts occurs in the form of filaments. The coking and die-off of zeolite catalysts which show some unique features in this respect, are discussed in the chapter on zeolite catalysts by Spencer.

The precursors to coke have been the subject of much discussion. Some work[63,64] has suggested that olefins are essentially responsible for coke deposition particularly at low conversions, whilst in other work[65] aromatics have been considered as the immediate precursors. Appleby et al.,[66] after studying the cracking of a wide range of hydrocarbons concluded that aromatic compounds are coke precursors, particularly those with condensed ring structures, e.g. anthracene.

Quantitative descriptions of coking began with empirical correlations, the best known of which was due to Voorhies[67] who found that the following relationship applied to a variety of catalysts, feedstocks, and flow rates in fixed and fluid bed reactors:

$$C = At^n \tag{10}$$

where C is the coke concentration on the catalyst, t is the time on line, A and n are constants.

Voorhies found that the coking rate in fixed bed operation was virtually independent of space velocity and the temperature dependence of the coking rate suggested an activation energy which was low and could indicate a diffusion controlled process. The latter view was reinforced by the fact that a number of workers[68] found a value for the exponent of 0.5. The general form of Voorhies' correlation fits a very large body of catalyst deactivation data, but it is only an empiricism and offers little or no insight into the possible mechanisms of coke formation.

More detailed experimentation by Eberly et al.,[65] whilst confirming the excellent correlations obtained with Voorhies' equation, found that the correlation constants are functions of reaction conditions. In particular, the correlation varied with feed rate and a variety of values of n between 0.5 and 1.0 were obtained. No dependence of coking on the particle size (in the range 75–2400 μm) was observed; evidence for the absence of diffusional effects.

Rudershausen & Watson,[69] in a study of the aromatization of cyclohexane over a molybdena–alumina catalyst, observed a strong dependence of the coking rate on temperature, indicating that their derived rate constants were related to the actual coking kinetics, and that there were again no diffusional effects. The kinetics, together with the observation that coke formation was suppressed by increasing the hydrogen partial pressure, led to a dual-site view of the coking mechanism. The indication is that coke is not formed by direct reaction on the cracking sites but by some other activity, the product of which is then strongly adsorbed onto the cracking sites.

The effect of coking on the mass transfer characteristics of cracking catalysts is structure-sensitive and individual to the system considered. For instance, the mass transfer characteristics (diffusivity) of a silica–alumina catalyst cracking ethylene were not affected by carbon deposition of up to 1%,[70] and Haldeman & Botty[61] also found little effect of coke deposit on surface area or pore size distribution at coke levels of several per cent. By contrast, losses of surface area of 20–30% have been observed for coke levels of a few per cent.[71] These discrepancies have been explained by Levinter et al.,[72] who observed a limiting degree of coke formation at which surface area loss became pronounced: this limiting amount of coke was much less than the total amount required to fill the pore space of the catalyst, suggesting that pore-blocking was preventing the use of all the catalyst internal area. The extent of pore-blocking depended strongly on catalyst properties and reaction conditions.

Distribution of coke in a fixed bed reactor was examined theoretically by Froment & Bischoff,[73] who showed that the coke profile would be descending in the case of a parallel coking reaction, but ascending for a series mechanism:

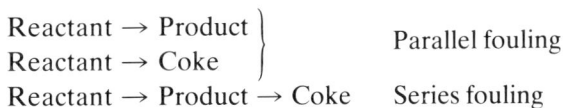

$$
\left.
\begin{array}{l}
\text{Reactant} \rightarrow \text{Product} \\
\text{Reactant} \rightarrow \text{Coke}
\end{array}
\right\} \qquad \text{Parallel fouling}
$$

$$
\text{Reactant} \rightarrow \text{Product} \rightarrow \text{Coke} \qquad \text{Series fouling}
$$

The distribution of coke in a single pellet is affected by diffusion limitations as well as whether the fouling process is parallel or series. In a parallel fouling process with no significant diffusion limitation, coking is uniform throughout the pellet; with a diffusion limitation an outer coked zone and an inner coke-free zone result. In a series fouling process with no significant diffusion effects, coking will occur in the centre of the pellet and spread outwards; with a diffusion limitation the reverse will occur and coke will only deposit in the outer layers of the pellet. A clear experimental demonstration of these effects was given by Murakami et al.,[74] using the reaction of toluene disproportionation to benzene and xylene (parallel coking) and the dehydrogenation of alcohols (series coking).

5.2   Fouling by impurity deposition

Because of the industrial importance of hydrocarbon processes, all of which are possibly subject to coking, it is the fouling of catalysts by carbon deposition which has received the most detailed consideration in the literature. There are other types of fouling reactions where deposition on the catalyst is caused by impurities and not from reactions intrinsically associated with the main chemical reactions, as is the case in coking. This type of fouling is found mainly in hydro-desulphurization reactors where petroleum residuals are hydrotreated to remove sulphur. Organometallic constituents react with hydrogen sulphide to produce solid deposits of metal sulphides. Liquefied coal extracts produce similar fouling problems when they are hydrotreated.[75] In petroleum residue the major metallic impurities are nickel and vanadium, and for coal-derived materials, nickel and iron are most important. The metal sulphide deposition occurs either in the catalyst pellet pores (pore-plugging) or between the pellets (bed-plugging). Interestingly, it has been shown[76] that the desulphurization reaction rate is greatest on small pore size catalysts (< 10 nm), whereas the demetallation reaction rate peaks at a pore size of $12-14$ nm. Methods of modelling the bed-plugging behaviour of these deposits have been published by Newson[60] and methods of minimizing the effect are discussed including grading the catalyst properties of activity and porosity through the catalyst bed. The deactivation of hydro-desulphurization catalysts has been reviewed elsewhere.[77]

Another unusual example of fouling by impurities is the deactivation of cobalt oxide ammonia oxidation catalysts.[6] In this case the deactivation was manifested by a fall in selectivity to nitric oxide over several months on line. The loss in oxidation selectivity is caused by a fall in the exposed surface area of cobalt oxide, chiefly due to fouling by impurities. The impurities in this case were not present in the reactant stream, but were present in the bulk of the catalyst originally and migrated to the catalyst surface during operation. The impurities were typically calcium and lead (as evidenced by ESCA studies) and the rate of migration increased as the operating temperature was raised. The deactivated catalyst could be regenerated by taking it off line and subjecting it to a dilute nitric acid wash which removed these impurity foulants from the surface. However, during further operation at reaction temperatures (*c.* 800°C) the process of what may be termed self-fouling continued and limited both the effective life of the catalyst and the conditions under which it could be operated.

6   Mass transfer effects

In practical situations catalysts are generally in the form of porous shapes, for instance, pellets, granules or extrusions. In the situation of gases flowing past these solid catalytic shapes there is always the possibility of a mass transfer limitation (and an associated heat transfer limitation) either externally to the pellet or within the pellet pores. Such resistances may occur singly or combined and may effectively reduce the observed reaction rate. Rates of mass or heat transfer have a much lower temperature coefficient than chemical reaction rates, so the physical processes tend to become rate-limiting at higher temperatures. A complicating factor in diffusion in the pores of catalyst pellets is the general lack of detailed knowledge of the pore structure. The problem of diffusion limitation in porous catalyst pellets has been extensively examined.[78-80] An important generalization for the physical limitations in catalyst systems was first noted by Carberry,[81] that for usual values of operating parameters the major heat transfer resistance is external to the pellet while for mass transfer the major resistance resides within the porous catalyst. Extensive reviews of the effects of mass and heat transfer resistances on reacting catalysts have been given by Satterfield[82] and Carberry.[83]

6.1   Diffusion and deactivation

So far, factors and changes which reduce the activity of catalyst surfaces have been described and classified. Just as heat and mass transfer limitations modify the observed performance of non-deactivating catalysts, so when deactivating changes occur in porous catalyst pellets their effects are modified by possible mass and heat transfer limitations. For example, if a reaction is operating under a severe external mass transfer limitation, then some initial deactivation (e.g. by poisoning or sintering) can occur without any observable decrease in the measured reaction rate. Such a situation pertains in the oxidation of ammonia to nitric oxide,[6] where deactivation of the catalyst first exhibits as fall in apparent selectivity of the oxidation to nitric oxide. Loss in catalyst activity causes an increased partial pressure of ammonia in contact with the catalyst surface as the reaction becomes more chemically limited and less mass transfer limited. The reaction order of the side reaction to N_2 is higher than that of the desired reaction to nitric oxide, so lower catalyst activity produces lower apparent selectivity.

The problem of a porous catalyst within which a first order reaction and a poisoning reaction were occurring was first considered by Wheeler.[84] Two types of deactivation may be distinguished, termed uniform and non-uniform. Either the rate of poison deposition may be low relative to its rate of transport in the pores, in which case the catalyst will be deactivated evenly (*uniform deactivation*), or the rate of poison deposition may be fast relative to its transport rate and the catalyst will deactivate preferentially at the pore mouth (*non-uniform* or *pore mouth deactivation*). Catalytic activity of the pellet is affected in different ways as a function of the level of poisoning in these two cases, depending on the extent of transport limitations on the catalytic reaction itself. Clearly, if the main reaction is slow and unaffected by pore diffusion limitations, the effective rate will simply fall in proportion to the level of poisoning whether uniformly or non-uniformly deposited (see Fig. 1.3). If the main reaction is pore diffusion limited and the poisoning is pore mouth poisoning, the effective catalyst activity falls to very low levels even when the level of deactivation is only a small fraction of the total catalyst surface. The reason for such an extreme effect (see Fig. 1.3) is that the outer layer of the particle which would offer the surface area used for reaction is precisely that which is rendered inactive by pore mouth deactivation.

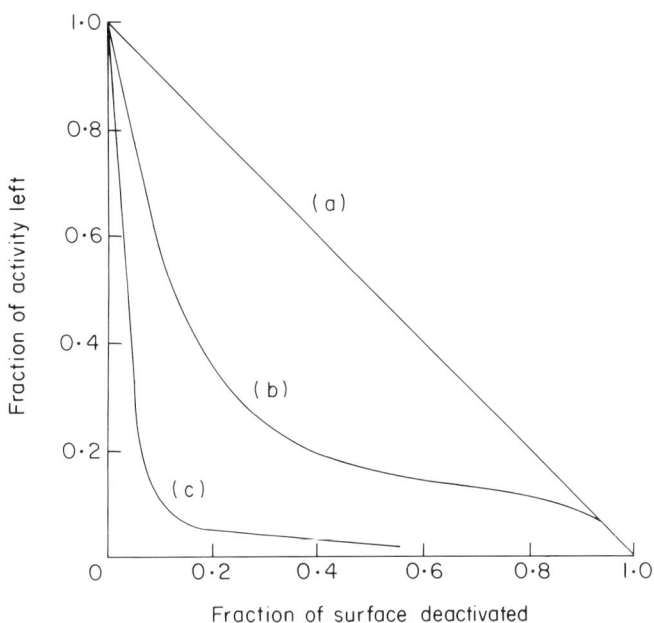

Fig. 1.3. Deactivation in a porous catalyst. (a) Any poison distribution; catalytic reaction not diffusion limited. (b) Pore mouth poisoning; catalytic reaction pore diffusion limited. (c) As (b) but more severe limitation. (After Wheeler, reference 84.)

The apparent activation energy for the main reaction is severely distorted in diffusion limited reactions when pore mouth deactivation is present.[84] Following curve C in Fig. 1.4 from the low temperature regime, initially the reaction rate is so slow that the unpoisoned surface is completely available to reaction and diffusion though the deactivated pore mouth is not rate limiting. Here the true intrinsic activation energy is observed. As the temperature is raised the diffusion rate through the deactivated pore mouths will become too slow to support the fast reaction on the unpoisoned part of the catalyst and the apparent activation energy will start to fall. Finally, at sufficiently high temperatures the catalytic rate is so high that it is completely controlled by the diffusion rate through the deactivated pore mouths and the activation energy falls to practically zero (a few $kJ\,mol^{-1}$; the

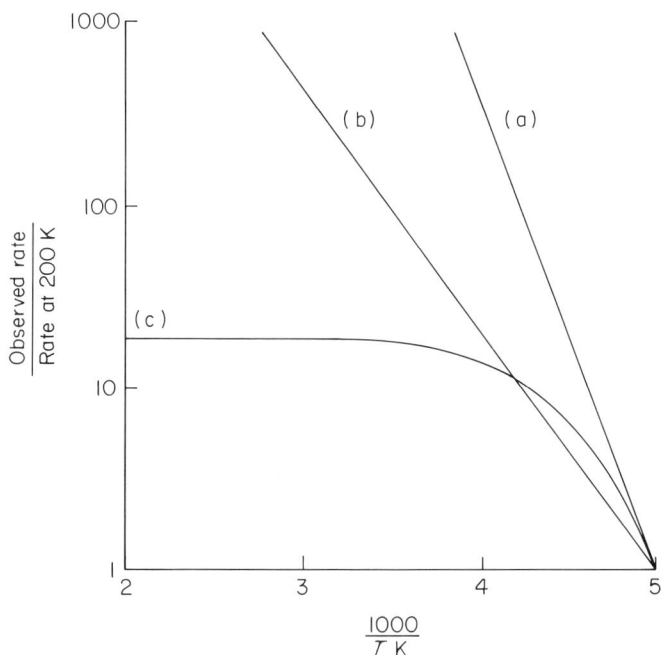

Fig. 1.4. Effect of poisoning on apparent activation energy. Hypothetical case (after Wheeler, reference 84): catalytic reaction with 46 kJ intrinsic activation energy. (a) No poison; no pore diffusion limitation, $E = 46$ kJ mol^{-1}. (b) No poison; pore diffusion limitation, $E_A \simeq E/2$. (c) Poisoned to 50% loss of surface (pore mouth poisoning) and pore diffusion limitation.

usual value for a mass-transfer process). There is then pore diffusion control of the reaction, not merely limitation.

6.2   General analyses of intraparticle deactivation processes

Somewhat more general analyses of intraparticle deactivation effects have been given by Masamune & Smith[85] and Murakami *et al.*[74] The analysis of Masamune & Smith considered the fouling of catalyst pellets by parallel and series mechanisms and also the case of fouling by an impurity in the feed. The assumptions used in their model included those of isothermality, the absence of external film diffusion resistances, and linear kinetics of deactivation in poison or coke concentration on the surface. Their results were expressed as diagrams showing profiles of both activity and concentration of reactants and products across the pellet radius as a function of elapsed time. Typically for intermediate values of the Thiele modulus ($\phi = 5$), the series mechanism gives maximum product and therefore maximum coke and minimum activity in the centre of the pellet, with profiles which become flatter as time goes on. The parallel mechanism produces maximum coke at the edges of the pellet where the reactant is maximum and so the outside of the pellet

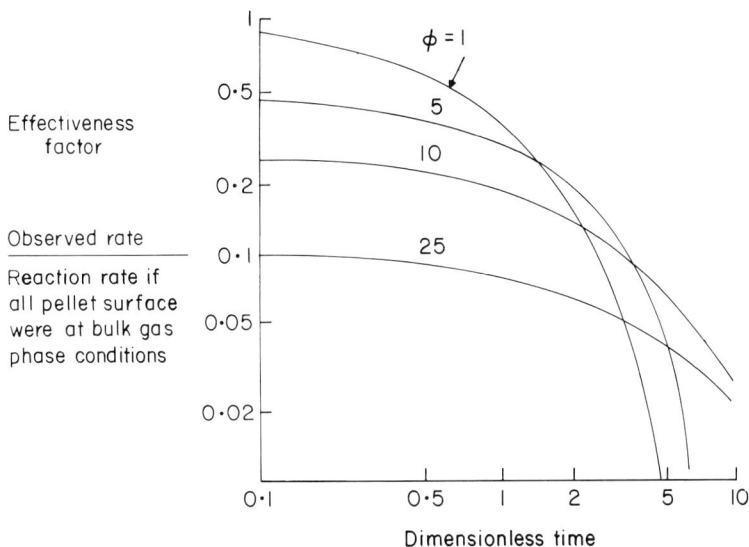

Fig. 1.5. Parallel fouling (after Masamune & Smith, reference 85).

is most heavily deactivated. Again the profiles become flatter with time. An interesting feature of the parallel mechanism is shown in Fig. 1.5, where the catalyst effectiveness factor is plotted against time for different degrees of diffusion limitation (Thiele modulus $\phi = 1$ to 25). There is a cross-over of curves for different values of the Thiele modulus and after a long period of operation the pellet which was initially the most diffusion limited becomes the most active. This isothermal analysis was extended by Murakami et al.[74] to include cases where external film resistances exist, and where a significant portion of the feed is converted to coke. The interesting feature in their results for the series mechanism of a coke profile reversal with change in Thiele modulus, has already been described (section 5.1). The effect of external film resistance was not very significant. Further exhaustive modelling studies of coking processes on catalyst pellets have been made by Kam & Hughes and co-workers,[86,87] covering non-isothermal behaviour, film effects, transient analysis and a full Langmuir–Hinshelwood analysis of fouling.

As already mentioned, Masamune & Smith[85] also considered the case of fouling/poisoning by an impurity in the feed. In this case two different Thiele moduli have to be considered, one for the catalytic reaction (ϕ) and one for the poisoning reaction (ϕ_p). The effectiveness factors obtained showed a decrease with time on stream, as would be expected, the decrease being markedly greater with increasing Thiele modulus for the catalytic reaction (Fig. 1.6). Least de-activation occurs in the case of minimum diffusion resistance for the catalytic reaction and a maximum resistance to diffusion for the poisoning reaction. More detailed analyses of irreversible impurity poisoning have been given by Hegedus[88] and Hegedus & McCabe.[89] Analyses of reversible poisoning have been given by Gioia[90] and Valdman et al.[91] Reversible poisoning is important in some industrial processes, chiefly as a means of controlling one or more of the main reactions. It was observed in these analyses that the time required to recover the original catalyst activity when the impurity was removed from the feed was much greater than the time required for poisoning. Results obtained by Valdman for both continuous and intermittent poisoning showed that pellets having an intermediate value for Thiele modulus (c. 2) often showed the highest productivity (area under the effectiveness factor against time curve).

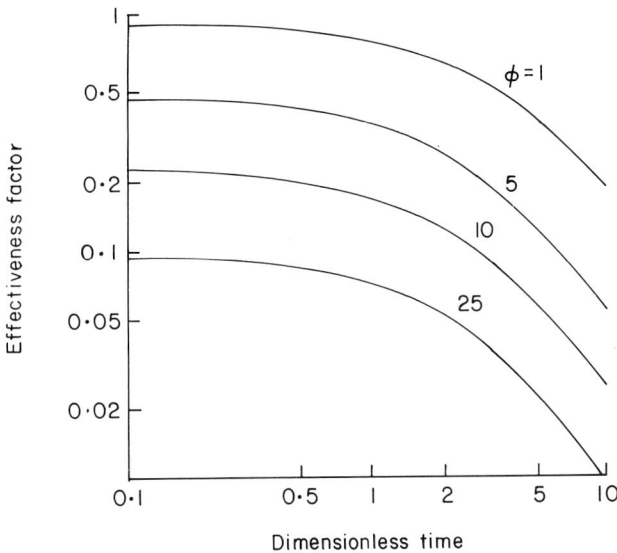

Fig. 1.6. Independent fouling/poisoning (as Fig. 1.5). ϕ_p (poisoning reaction) = 10.

7   Deactivation in catalyst reactors

So far the processes which cause changes in catalytic surfaces leading to a loss of catalytic activity have been classified and discussed. The way in which mass transfer effects influence the observed effect of these surface changes in real catalyst pellets was then referred to. Finally, since all real catalysts are operated in reactors, often containing many tons of catalyst, it is apposite to consider briefly deactivation in catalytic reactors. Some reactors (fluidized-bed reactors, moving-bed reactors) are designed to receive transport of solid catalyst and can readily have continuous catalyst regeneration or replacement facilities incorporated into the reactor system. This is not the case with fixed-bed reactors, a common design in which the gas stream passes through a bed of solid catalyst particles. They are complicated physically and chemically and their performance is not easy to predict accurately. Even though many industrial reactors approximate to plug flow, temperature gradients in most practical reactors exert a large influence on their performance. In addition, heat and mass transfer effects between gas and solid phases have to be built into any reactor models. Division of deactivation processes into classes is less important when a whole reactor is considered, as reactor performance and the position of deactivated areas become similar for many classes of deactivation.

The first detailed discussion of the problem of impurity poisoning in fixed-bed reactors was given by Wheeler & Robell.[92] They computed the conversion in an isothermal plug flow reactor for a first order reaction where the rate constant was a function of poison concentration (as per Wheeler's poisoning curves).[84] The distribution of poison in the bed was determined from the fixed-bed adsorption theory of Bohart & Adams.[93] The poison profiles obtained showed the development of a 'poison wave' in the bed which passed through the reactor at a fixed velocity and in a fixed shape. Experimental confirmation was obtained by a study of the H_2S poisoning of a Pt/Pd on alumina catalyst for the oxidation of CO. An extension of their analysis was made by Haynes,[94] who solved the system for cases where the poisoning reaction as well as the main reaction were diffusion-limited using separate Thiele module for reactants and poison.

Both of these studies were for isothermal systems but the majority of industrial systems are not isothermal, often having steep temperature gradients in the reactor. Commonly, in an adiabatic reactor the reaction is confined to a narrow zone in the bed—as evidenced by the steep temperature rise—and as poisoning occurs this steep temperature profile moves down the bed with little change in shape. Such poisoning behaviour is shown by many catalysts, e.g. water gas shift catalysts and methanation catalysts. One detailed study of a non-isothermal system[95] considered the thiophene poisoning of a nickel/kieselguhr catalyst for the hydrogenation of benzene, and concluded that poisoning kinetics may be more complex than generally supposed.

There have been a number of studies of coking in fixed-bed reactors, the first detailed analyses of isothermal coking being due to Froment & Bischoff.[73,96] Calculated carbon profiles showed the greatest deposition of coke at the entrance of the reactor for parallel fouling and greatest near the exit for series fouling. Ervin & Luss studied non-isothermal fouling in fixed-bed reactors[97] and employed a full transient analysis but assumed first order kinetics for the main reaction and neglected intraparticle concentration gradients. Kam & Hughes[98] modelled adiabatic fixed-bed reactors using Langmuir–Hinshelwood kinetics for the main reaction and including intraparticle gradients. It was shown that the activation energy parameter (E/RT gas) for fouling was an important parameter and, as expected with adiabatic operation for an exothermic main reaction, more deactivation occurred as that parameter increased. This analysis was extended by Brito-Alayon et al.[99] to include Langmuir–Hinshelwood kinetics for the coking reaction and showed that for parallel fouling under exothermic conditions Langmuir–Hinshelwood fouling could give contradictory behaviour to what may be called normal parallel fouling behaviour. The effect of diffusional resistance on activity profiles was also determined and it was shown that when Langmuir–Hinshelwood kinetics apply, such resistance can have profound influences on the shape of the activity profile.

Finally, some discussion of the influence of catalyst deactivation on fixed-bed reactor dynamics has appeared in the literature, concerning reactor start-up[100] and thermal factors in the reactor,[101] but many such features remain unresolved.

8   References

1   Lieber, E. & Morritz, F.L. *Adv. Cat.* 1953, **5**, 417.
2   Friedrich, J.B., Wainwright, M.S. & Young, D.J. *J. Catal.* 1983, **80**, 1, 14.
3   US patent 4181630, 1978.
4   Atkinson, G.B., Nicks, L.J., Baglin, E.G. & Bauer, D.J. *US Bureau of Mines* R. of I. 8631, 1982.
5   Hughes, R. *Deactivation of Catalysts.* Academic Press, New York, 1984.
6   Andrew, S.P.S. & Chinchen, G.C. *Catalyst Deactivation*, Delmon & Froment (Eds). Elsevier, Amsterdam, 1980.
7   Hegedus, L.L. & McCabe, R.W. *Int. Symposium on Catalyst Deactivation.* Antwerp, Belgium, 1980.
8   Berkman, S., Morrell, J.C. & Egloff, G. *Catalysis.* Reinhold, New York, 1940.
9   Innes, W.B. *Catalysis*, Emmett, P.H. (Ed.), Vol. 1. Reinhold, New York, 1954.
10   Knozinger, H. *Adv. in Catalysis*, 1976, **25**, 184.
11   Butt, J.B. *Adv. Chem. Ser.* 1972, **109**, 259.
12   Butt, J.B. *Int. Symposium on Catalyst Deactivation.* Antwerp, Belgium, 1980.
13   Petro, J. *Contact Catalysis*, Szabo & Kallo (Eds). Elsevier, New York, 1976.
14   Levenspiel, O. *Chemical Reaction Engineering*, Ch.15. John Wiley, New York, 1972.
15   Pashley, D.W., Stowell, M.J., Jacobs, M.H. & Law, T.J. Phil. Mag. 1964, **10**, 127.
16   Coble, R.L. *J. Am. Ceram. Soc.* 1958, **41**, 55.
17   Hoogschagen, J. & Zwietering, P. *J. Chem. Phys.* 1953, **21**, 2224.
18   Chinchen, G.C., Logan, R.H. & Spencer, M.S. *Applied Catalysis*, 1984, **12**, 89.
19   Williams, A., Butler, G.A. & Hammonds, J. *J. Catal.* 1972, **24**, 352.
20   Hermann, R.A., Adler, S.F., Goldstein, M.S. & DeBaun, R.H. *J. Phys. Chem.* 1961, **65**, 2189.
21   Hughes, T.R., Houston, R.J. & Sieg, R.P. *Ind. Eng. Chem.* (Proc. Des. Devel.) 1962, **1**, 96.
22   Gruber, H.L. *J. Phys. Chem.* 1962, **66**, 48.
23   Maat, H.J. & Moscou, L. *Proc. 3rd Int. Congr. Catalysis*, 1965, p. 1277. North-Holland, Amsterdam.
24   Flynn, P.E. & Wanke, S.E. *Catal. Rev. Sci. Eng.* 1975, **12**, 93.
25   Ruckenstein, E. & Pulvermacher, B. *A.I.Ch.E.J.* 1973, **19**, 356.
26   Ruckenstein, E. & Pulvermacher, B. *J. Catal.* 1973, **29**, 224.
27   Flynn, P.C. & Wanke, S.E. *J. Catal.* 1974, **34**, 390.
28   Flynn, P.C. & Wanke, S.E. *J. Catal.* 1974, **34**, 400.
29   Wynblatt, P. & Gjostein, N.A. *Progr. Solid State Chem.* 1975, **9**, 21.
30   Wanke, S.E. *J. Catal.* 1977, **46**, 234.

31   Granqvist, C.G. & Burhman, R.A. *J. Catal.* 1977, **46**, 238.
32   Ruckenstein, E. & Dadyburjor, D.B. *J. Catal.* 1977, **48**, 73.
33   Bassett, G.A. *Proc. Int. Symp. on Condensation and Evaporation of Solids.* Gordon & Beach, New York, 1964.
34   Skofronick, J.G. & Phillips, W.R. *J. Appl. Phys.* 1967, **38**, 4791.
35   Granqvist, C.G. & Burhman, R.A. *J. Appl. Phys.* 1975, **27**, 693. *Solid State Commun.* 1976, **17**, 123. *J. App. Phys.* 1976, **47**, 2200.
36   Granqvist, C.G. & Burhman, R.A. *J. Catal.* 1976, **42**, 477.
37   Richardson, J.T. & Crump, J.G. *J. Catal.* 1979, **57**, 417.
38   Nakamura, M. *J. Catal.* 1975, **39**, 125.
39   Bett, J.A., Kinoshita, A. & Stonehart, P. *J. Catal.* 1974, **35**, 307.
40   Richardson, J.T. & Desai, P. *J. Catal.* 1976, **42**, 294.
41   Kuo, H.K., Ganesan, P. & DeAngeus, R.J. *J. Catal.* 1980, **64**, 303.
42   Campbell, J.S. *Ind. Eng. Chem.* (Proc. Des. Develop.) 1970, **9**, 588.
43   Delamare, F. & Rhead, G.E. *Compt. Rend.* 1970, **270**, 249.
44   Clay, R.D. & Petersen, E.E. *J. Catal.* 1970, **16**, 32.
45   Andrew, S.P.S. *Catalyst Handbook.* Wolfe Scientific Books, London, 1970.
46   Chinchen, G.C. *Proc. Fertiliser Soc.* 1978, **171**.
47   Gioia, F. *Ind. Eng. Chem. (Fund)* 1971, **10**, 204.
48   Maxted, E.B. *Adv. in Catalysis*, 1951, **3**, 129.
49   Mills, G.A., Boedeker, E.R. & Oblad, A.G. *J. Amer. Chem. Soc.* 1950, **72**, 1554.
50   Apelbaum, L.O., Berezina, Yu. I. & Témkin, M.I. *Russian J. Phys. Chem.* 1960, **34**, 1314.
51   Chinchen, G.C. Unpublished work.
52   Chinchen, G.C. *Catalyst Handbook.* Wolfe Scientific Books, London, 1970.
53   Morita, S. & Inoue, T. *Int. Chem. Eng.* 1965, **5**, No. 1, 180.
54   Clay, R.D. & Petersen, E.E. *J. Catal.* 1970, **16**, 32.
55   Pease, R.N. & Stewart, L. *J. Amer. Chem. Soc.* 1925, **47**, 1235.
56   Szepe, S. & Levenspiel, O. *Proc. European Fed., 4th Chem. Reaction Eng.*, Brussels. Pergamon Press, Oxford, 1970.
57   Levenspiel, O. *et al. J. Catal.* 1972, **25**, 265.
58   Butt, J.B., Wachter, C.K. & Billimoria, R.M. *Chem. Eng. Sci.* 1978, **33**, 1321.
59   Petersen, E.E. & Pachelo, M.A. *ACS Symp. Ser.* 1984, **237**, 363.
60   Newson, E.J. *Ind. Eng. Chem. (Proc. Des. Devel.)* 1975, **14**, 27.
61   Haldeman, R.G. & Botty, M.C. *J. Phys. Chem.* 1959, **63**, 489.
62   Trimm, D.L. *Catal. Rev. Sci. Eng.* 1977, **16**, 135.
63   Wojciechowski, B.W. *Catal. Rev. Sci. Eng.* 1974, **9**, 79.
64   Blue, R.W. & Engle, C.J. *Ind. Eng. Chem.* 1951, **43**, 494.
65   Eberly, P.E., Kimberlin, C.N., Miller, W.H. & Drushel, H.V. *Ind. Eng. Chem. (Proc. Des. Devel.)* 1966, **5**, 193.
66   Appleby, W.G., Gibson, J.W. & Good, G.M. *Ind. Eng. Chem. (Proc. Des. Devel.)* 1962, **1**, 102.
67   Voorhies, A. *Ind. Eng. Chem.* 1945, **37**, 318.
68   Prater, C.D. & Lago, R.M. *Adv. Catal.* 1956, **8**, 293.
69   Rudershausen, C.G. & Watson, C.C. *Chem. Eng. Sci.* 1955, **3**, 110.
70   Ozawa, Y. & Bischoff, K.B. *Ind. Eng. Chem. (Proc. Des. Devel.)* 1968, **7**, 67.
71   Ramser, J.N. & Hill, P.B. *Ind. Eng. Chem.* 1958, **50**, 117.
72   Levinter, M.E., Manchenkov, G.M. & Tanatarov, M.A. *Int. Chem. Eng.* 1967, **7**, 23.
73   Froment, G.F. & Bischoff, K.B. *Chem. Eng. Sci.* 1961, **16**, 189.
74   Murakami, Y., Kobayashi, T., Hattori, T. & Masuda, M. *Ind. Eng. Chem. (Fund)* 1968, **7**, 599.
75   Davies, G.O. *Chem. and Ind.* 1978, 560.
76   Inoguchi, M. *Skokubai*, 1976, **18** (3), 78.
77   Ohtsuka, T. *Catal. Rev. Sci. Eng.* 1977, **16**, 291.
78   Damkohler, G. *Z. Phys. Chem.* A193, 1943, **16**.

79   Thiele, E.W. *Ind. Eng. Chem.* 1939, **31**, 916.
80   Wheeler, A. *Adv. Catal.* 1951, **3**, 250.
81   Carberry, J.J. *Ind. Eng. Chem.* 1966, **58**, 40.
82   Satterfield, C.N. *Mass Transfer in Heterogeneous Catalysis.* MIT Press, Cambridge, 1970.
83   Carberry, J.J. *Chemical and Catalytic Reaction Engineering.* McGraw-Hill, New York, 1976.
84   Wheeler, A. *Catalysis,* Emmet, P.H. (Ed.), Vol. 2. Reinhold, New York, 1955.
85   Masamune, S. & Smith, J.M. *Am. Inst. Chem. Eng. J.* 1966, **12**, 384.
86   Kam, E.K.T., Ramachandran, P.A. & Hughes, R. *J. Catal.* 1975, **38**, 283. *Chem. Eng. Sci.* 1977, **32**, 1307, 1317.
87   Kam, E.K.T. & Hughes, R. *A.I.Ch.E.J.* 1979, **25**, 359.
88   Hegedus, L. *Ind. Eng. Chem. (Fund)* 1974, **13**, 190.
89   Hegedus, L. & McCabe, R.W. *Cat. Rev. Sci. Eng.* 1981, **23**, No. 3, 377.
90   Gioia, F. *Ind. Eng. Chem. (Fund)* 1971, **10**, 204.
91   Valdman, B., Ramachandran, P.A. & Hughes, R. *J. Catal.* 1976, **42**, 303.
92   Wheeler, A. & Robell, A.J. *J. Catal.* 1969, **13**, 299.
93   Bohart, G. & Adams, E. *J. Am. Chem. Soc.* 1920, **42**, 523.
94   Haynes, H.W. *Chem. Eng. Sci.* 1970, **25**, 1615.
95   Weng, H.S., Eigenberger, G. & Butt, J.B. *Chem. Eng. Sci.* 1976, **31**, 1341.
96   Froment, G.F. & Bischoff, K.B. *Chem. Eng. Sci.* 1962, **17**, 105.
97   Ervin, M.A. & Luss, D. *A.I.Ch.E.J.* 1970, **16**, 979.
98   Kam, E.K.T. & Hughes, R. *Chem. Eng. J.* 1979, **18**, 93.
99   Brito-Alayon, A., Hughes, R. & Kam, E.K.T. *Chem. Eng. Sci.* 1981, **36**, 445. *Chem. Eng. J.* 1982, **24**, 123.
100 Blaum, E. *Chem. Eng. Sci.* 1974, **29**, 2263.
101 Billimoria, R.B. & Butt, J.B. *Chem. Eng. J.* 1981, **22**, 71.

2 Applications of physical techniques in heterogeneous catalysis

J.S. Foord

1   Introduction

Although heterogeneous catalysis has long been a subject of huge industrial importance and scientific interest, the preparation of the catalysts themselves has remained until recently an essentially empirical art. This indicates some of the difficulties and complexities associated with attempts to characterize the physico-chemical properties of catalysts; such information is required if progress is to be made in the direction of structured catalyst design.[1] Since catalytic action takes place at specific sites on solid surfaces, the physical structure of industrial catalysts is engineered to maximize the surface area of the active phase. Most commonly, this is achieved by dispersing the catalytic material on high surface area 'inert' supports, such as alumina or silica.[2] These materials have surface areas in the range 1–1000 m^2 g^{-1}, the overwhelming fraction of which is associated with accessible internal pores in the support particles. Dispersed catalyst particles are located on the external and internal surfaces of the supporting material and typically range in size between 1 and 50 nm. Achieving a full understanding of the reactive properties of catalysts so formulated is indeed a formidable problem. The identity of the active surfaces and sites which constitute the *working* catalyst must be established. Even in the case of catalysts formed from single metals this is not straightforward; additional chemicals ('promoters' and 'moderators') are often present and surface deposits (e.g. carbide) can build up during catalytic action. When catalysts such as mixed transition metal oxides containing, e.g. ten differing elements are considered, this problem becomes increasingly difficult.

The relevant surface reaction mechanisms governing conversion of reactants to products must be understood. A problem here is the need to make detailed *in situ* measurements on active catalysts under the extremes of temperature and pressure at which catalytic processes run. Most importantly, knowledge is required of the manner by which the physical structure of the catalyst at a molecular level affects its performance. Many variables may be involved. Reactivity can depend on particle size and morphology and interactions between catalyst and support need to be considered. Limitations on the diffusion of species through the pore structure of the catalyst influence selectivity and activity. Catalysts possess a dynamic structure, which varies according to pre-treatment and time on-stream, and this adds further complications.

The most traditional approach to catalyst characterization is the determination of reaction kinetics. The philosophy here is to treat the catalyst as a 'black-box'—by correlating rate and selectivity data with feedstock composition, catalyst constitution, and other process variables, it is hoped to understand how the catalyst functions. Such an approach unfortunately possesses severe limitations. Deduction of actual reaction mechanisms is inhibited because differing kinetic models often yield similar mathematical forms for the rate equation. Variations in catalyst preparation procedures may simultaneously alter a number of physicochemical properties and it is thus unclear exactly what is responsible for observed changes in reaction rate. While the measurement of catalyst kinetics represents an important step to the process engineer, it presents insufficient insight into the catalytic chemistry involved, both from the academic viewpoint of obtaining a fundamental understanding of catalysis and from the industrial viewpoint, where the optimization of catalytic formulations is the ultimate aim.

Much emphasis has consequently been placed in recent years on the development and application of physical techniques to study catalysts and catalytic processes at a molecular level. The goal is to obtain a detailed atomic picture of the structure of active catalysts and establish the relationship with the elementary chemical reactions responsible for overall catalytic transformations. The highly dispersed and disordered state in which catalysts exist limits the range of physical techniques which can be brought to bear. For example, although powder X-ray diffraction is useful for investigating the structure of bulk materials, when applied to the study of catalysts the reflections can be hopelessly broad and weak, because of the small crystallite size. Nevertheless, a substantial number of physical techniques are now available for the characterization of catalysts. Many of such methods can be used *in situ* under real process conditions; others are unlikely to be ever used under such conditions but are nonetheless still capable of imparting a good deal of useful information. The purpose of this article is to survey some of the techniques employed and describe their uses and capabilities. Rather than concentrating on the more established methods such as surface area measurement, radio-isotope exchange and differential scanning calorimetry, attention is largely devoted to the newer particle probes, which are currently undergoing rapid development and deployment.

The review is divided into two parts. The following section focuses on techniques applicable to the characterization of 'real' catalytic systems (such as may be encountered in large-scale industrial processes). The subsequent section then covers the large body of research on highly idealized model systems, which has made such a large contribution to our basic understanding of the nature of heterogeneous catalysis.

2   Techniques for the study of dispersed catalysts

2.1   *Extended X-ray absorption fine structure (EXAFS) measurements*

EXAFS has recently emerged as a most powerful tool for the structural investigation of dispersed catalyst systems. From an analysis of EXAFS data, one can obtain information on the number and type of neighbouring atoms about a given absorber atom, the interatomic distances and some estimate of local disorder (both thermally induced and otherwise). By obtaining EXAFS measurements for the different chemical constituents, such knowledge can be used to build up a detailed picture of catalyst structure. Since EXAFS is essentially a probe of local geometry, it is particularly suited to the study of heterogeneous catalysts, where long-range order is normally absent. Furthermore, the experiments can be carried out under controlled atmospheric conditions, providing opportunities for *in situ* investigations.

Detailed accounts of the EXAFS experiment are given elsewhere.[3,4] In brief, the X-ray photon absorption cross-section (μ) of the sample is measured as a function of photon energy. When the X-ray energy is just sufficient to cause emission of photoelectrons from an electronic core level, μ shows a sharp increase, as can be seen in Fig. 2.1. In the case of isolated atoms, μ exhibits a smooth fall-off as the photon energy increases beyond the edge. In all other cases an oscillatory variation ('absorption fine structure') in μ is observed, extending for ~ 1000 eV. This is again clearly visible in Fig. 2.1. Although the origin of the EXAFS was long a subject of debate, it is now clear that it relates to the local structure around the absorbing atom. The oscillation in μ arises from constructive or destructive interference of the emitted photoelectron wave with its reflection from surrounding atoms. EXAFS normally refers to the absorption fine structure

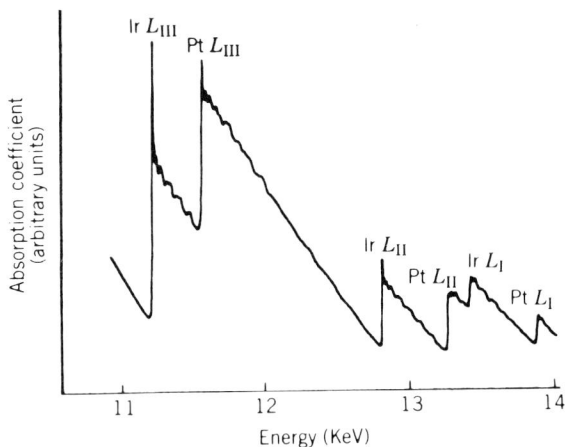

Fig. 2.1. X-ray absorption spectrum in the region of the L absorption edges of iridium and platinum for a platinum iridium, silica supported catalyst.[5] (Reprinted with permission from the American Institute of Physics.)

observed well away (> 50 eV) from the edge itself, where the theoretical analysis is relatively straightforward. Even stronger variations (XANES; 'X-ray absorption near edge structure') in μ may be visible at smaller photoelectron energies. Unfortunately, the theoretical description of XANES is a good deal more complex since a full multiple scattering calculation is required. XANES is sensitive to *bond angles* as well as bond distances and thus represents an area for important future developments.

EXAFS data are normally quantitatively represented as the variation in the EXAFS function $\chi(K)$ with photon energy or photoelectron wave vector K, where

$$\chi(K) = \frac{\mu - \mu_0}{\mu_0}$$

where μ refers to the measured atomic absorption cross-section and μ_0 to the value for a hypothetical free atom state. Analysis is normally carried out using computer programs incorporating a good theoretical description of the phenomenon, to produce a best fit to the measured data. A number of variable parameters enter the calculation and it is helpful if these can be derived or checked in experiments on relevant standard materials. Bond distances typically accurate to 0.002 nm are obtained; coordination numbers are rather less reliable ($\pm15\%$). For a rapid preliminary analysis and graphic data representation, the magnitude of the Fourier transform of a K-weighted EXAFS function is plotted out; the resulting graph exhibits peaks with appropriate intensities at the approximate near-neighbour distances in the sample.

Experimentally, an intense white source of X-ray radiation is the critical requirement. While the Bremsstrahlung from conventional rotating-anode X-ray sources may be used, measurement times are impractically long. EXAFS work is today mostly carried out using the much more intense synchrotron radiation sources. Catalytic samples are generally in the form of compressed pellets, 0.1–1 mm thick, which may be held in controlled environments, and the spectrum is recorded by measuring directly the extent of X-ray absorption or the fluorescence yield.

An interesting example of the application of EXAFS to catalyst characterization comes from the work of Joyner & Meehan[6] who have studied supported Pt catalysts. One question addressed concerned the structure of the supported metal particles; it was long ago suggested that microcrystallites of less than 150 atoms should adopt a fivefold symmetric icosahedral structure, rather than the normal close-packed arrangement.[7] A distinct difference between the two structures is the higher near-neighbour coordination number for the face-centred cubic (fcc) lattice and thus EXAFS provides a sensitive test of the theory. The EXAFS data clearly showed that the coordination numbers for the Pt in the supported catalysts were those expected for the conventional fcc structure. There is little reason therefore to suppose the existence of icosahedral structures. The same workers studied the reactivity of Pt/SiO$_2$ and Pt/C catalysts towards oxygen. After exposure to atmospheric pressure, the SiO$_2$ supported catalyst showed only the presence of Pt–O bonds, indicating that the metal particles had been completely oxidized. This suggests that the small particles are much more reactive towards oxygen than bulk Pt. In contrast, the Pt/C catalyst did not display this enhanced reactivity. Such differences in oxidation behaviour are intriguing but at present not understood.

EXAFS has been advantageously employed in the study of bimetallic catalysts, where the degree of mixing between the two metallic components is always a subject of central interest. Sinfelt and co-workers have carried out extensive studies of Ru–Cu, Os–Cu and Pt–Ir catalysts.[5,8,9] The immediate conclusion which can be taken directly from the recorded EXAFS spectra is that the metallic components in each of these catalysts exhibit a strong interaction with each other, since clear differences are observed between spectra recorded from the bimetallic catalysts and samples containing the individual metals. In the case of Ru–Cu (Os–Cu) catalysts, the authors were able to compute the separate contributions of nearest neighbour Ru(Os) and Cu back-scattering atoms to the recorded spectra above the Cu-K and Ru(Os) L absorption edges. The agreement between theory and experiment is excellent, as can be seen in Fig. 2.2, where typical data are presented. From such data, it was shown that the Cu is coordinated to essentially equal numbers of Ru(Os) and Cu atoms, while the Ru(Os) is largely only coordinated to Ru(Os). This indicates that the two metallic components are not

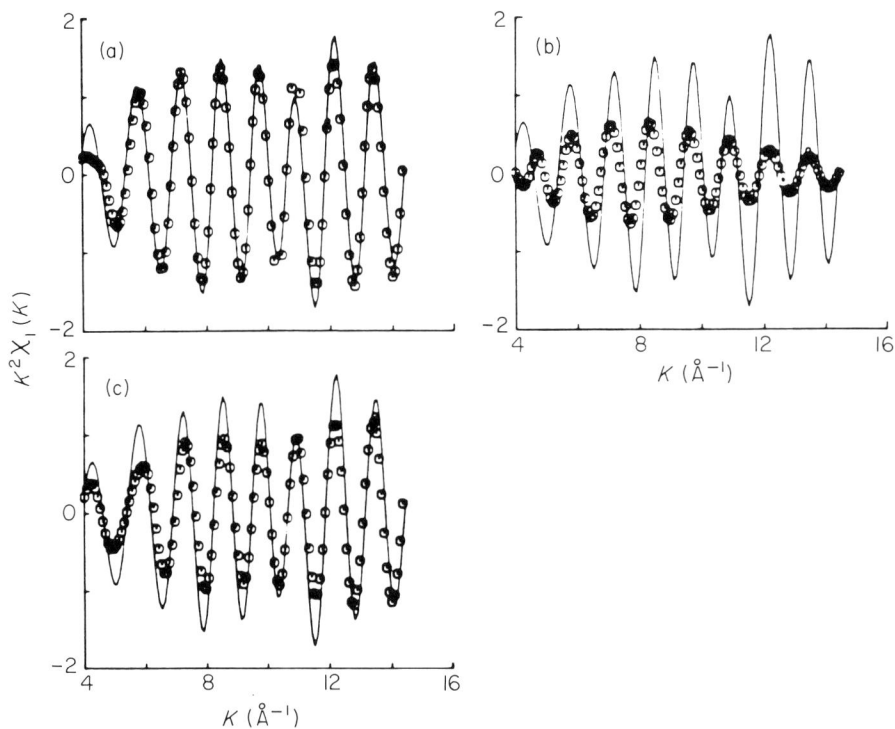

Fig. 2.2. Contributions of nearest neighbour Cu (field b) and Os (field c) to the EXAFS (field a), associated with the Os L_{III} absorption edge for a silica-supported Os–Cu catalyst.[9] Points are calculated and the solid lines are derived from experiment. (Reprinted with permission from the American Institute of Physics.)

uniformly distributed through the supported metal clusters. Upon exposure of the catalysts to oxygen, EXAFS indicated that the platinum metal component was less oxidized if Cu was present, whereas the Cu component became more oxidized in the presence of Ru or Os. Using such information in combination with other evidence, it was possible to deduce that the bimetallic particles consist of a core of Ru(Os) with small patches or multiplets of Cu distributed over the surface.

The application of EXAFS has also been extended to the study of metal compound catalysts. Co, Mo/Al$_2$O$_3$ dehydrosulphurization catalysts have been examined and the evolution in structure, which accompanies their conversion from the oxidic to the active sulphided form investigated.[10] EXAFS was able to cast light on which of the many possible Co$_x$Mo$_y$S$_z$O phases are present at the various stages of activation.

In an EXAFS investigation of anatase-supported V$_2$O$_5$ catalysts,[11] an explanation has been provided for the very strong support-enhancement which is exhibited by this system. Previous explanations of this classic enhancement effect have suggested the answer lies in the particular epitactic growth mode of crystalline V_2O_5 on the anatase support. However, the measured EXAFS and XANES data indicated this was incorrect since crystalline V_2O_5 was found to be absent from the catalyst. Instead, the predominant unit within the active phase was shown to consist of two double-bonded oxygen ligands attached to vanadium species, which are anchored to the support by means of V–O–Ti bridges. The structure of this unit varies with the nature of the support, hence the enhancement effect is explained.

Such studies tend to emphasize the static structure of catalysts. A fascinating insight into the dynamic changes which can occur comes from an EXAFS investigation of Rh/Al_2O_3 catalysts in a controlled CO environment.[12] It was demonstrated that metal crystallites are present within the catalyst in the absence of CO. However, in the presence of CO, the exothermic formation of Rh–CO bonds was found to cause complete disruption of the metal–metal bonding, leading ultimately to the atomic dispersion of Rh over the support. IR and electron microscope studies had long been at odds concerning the degree of metal dispersion within this catalyst. The reason for this disagreement is now clear—the two sets of measurements were carried out in very different atmospheric environments.

These examples demonstrate the great power of EXAFS when applied to catalyst characterization. The importance of the technique will continue to grow in the future, as EXAFS facilities become available to a greater number of workers and data analysis more routine. An obvious area of investigation will be the role played by catalyst promoters and it is expected that more emphasis will be placed on *in situ* studies of reaction kinetics. This will be greatly aided if the time taken to record the EXAFS data can be reduced. An important development here is the so-called 'energy-dispersive' EXAFS experiment, which permits the simultaneous recording of EXAFS over a spread of photon energies, thus greatly increasing data acquisition rates.

2.2   *Nuclear magnetic resonance spectroscopy*

Although NMR is a central technique for the investigation of molecular structure and fluxional behaviour of molecules either in solution, or in the liquid phase[13], its role as a tool for the characterization of solids is much less pronounced. This is principally because NMR spectra of the liquid state display rich structure from which a wealth of information on the compound under investigation may be gleaned, whereas corresponding spectra for the solid state may reveal broad featureless absorption envelopes. Such differences arise because molecules in liquids undergo rapid isotropic tumbling whereas in solids they are (more or less) fixed in space. This has a profound effect on measured line widths. Dipolar interactions between magnetic nuclei in close vicinity give rise to a large broadening of the absorption lines from solids, whereas in liquids the rapid molecular motion averages the interactions to zero and removes the broadening effect.

Secondly, nuclei in solids undergo a chemical shift which is dependent on the orientation to the applied magnetic field and this so-called chemical shift anisotropy makes a further contribution to the line broadening. Fortunately, developments over the past two decades now permit these broadening effects to be eliminated. With regard to chemical shift anisotropy, the effects can be removed by rapidly spinning the sample at the 'magic-angle' of 54.7° with respect to the external magnetic field, as first described by Andrew,[14] and Lowe.[15] Provided the rotation frequency is greater than the line-width of the spectrum before spinning, the chemical shift is reduced to its isotropic value and improved resolution results. The dipolar broadening is weak for samples where the concentration of magnetic nuclei is low (e.g. samples where ^{13}C is the only magnetic nucleus present) and in such instances the application of magic-angle spinning is sufficient to yield well resolved spectra. In problems of interest to catalysis, protons are often present and dipolar coupling is too large to be removed using sample spinning techniques. Heteronuclear coupling may be removed by strong continuous or pulsed irradiation at the proton resonance frequency while the NMR spectrum of the species of interest, e.g. ^{13}C is recorded. Homonuclear ^{1}H coupling can be removed by applying programmed pulse trains to the sample, thus also permitting well resolved spectra of ^{1}H nuclei to be observed in solids. With regard to surface studies, high surface area materials are required to obtain adequate signal–noise ratios but this is not a problem for catalytic work where such a condition is generally satisfied. In NMR studies of low abundance nuclei the signal–noise ratio may be enhanced using 'cross-polarization' involving the more abundant magnetic nuclei (usually ^{1}H). The various techniques of solid state NMR are described in detail elsewhere.[16] By using such techniques, information is obtained on both the catalyst structure and the nature of the adsorbed species which may be present.

The area in which solid state NMR has had the greatest single impact is the structural characterization of zeolite catalysts. These are in many ways ideal compounds to study with the technique. NMR spectra are obtainable for all of the constituents of the aluminosilicate framework (^{29}Si, ^{27}Al, ^{17}O), as well as the majority of the exchangeable cations. ^{1}H nuclei covalently bonded to the zeolite framework are absent and adequate resolution can be obtained in a simple magic-angle spinning (MAS) experiment. Distinct structural groups occur in the zeolite framework which are resolvable by measurement of chemical shifts. To date, ^{29}Si MASNMR has provided the most important information. There are five possible Si environments in zeolites, corresponding to $Si(OAl)_n(OSi)_{4-n}$ ($n = 0-4$) units. As shown first by Lippmaa & Engelhardt,[17] five peaks in the ^{29}Si NMR spectra may be observed corresponding to these units; the types of linkage present in the zeolite framework can thus be readily determined and questions pertaining to the Si–Al ordering addressed. The spectra typically obtainable are

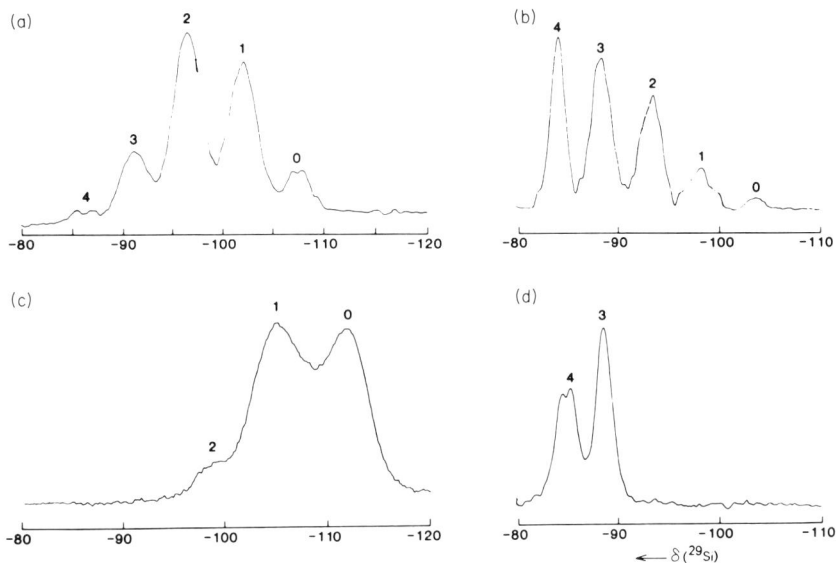

Fig. 2.3. ^{29}Si MASNMR spectra of (a) gmelinite, (b) analcite, (c) mordenite and (d) zeolite Z. Numbers above the peaks are the n in $Si(OAL)_n(OSi)_{4-n}$.[18] (Reprinted with permission from Verlag Chemie Gmbh.)

illustrated in Fig. 2.3. The ^{29}Si MASNMR spectrum of zeolite X when Si:Al = 1 consists of one peak at a chemical shift consistent with the presence of $Si(OAl)_4$ species; it thus follows that Si and Al undergo a strict alternation within the aluminosilicate structure, clearly demonstrating that Loewensteins rule, which states that two Al atoms cannot occupy adjacent tetrahedral sites, is obeyed.[17] Although ^{29}Si MASNMR is probably most helpful, information is also obtainable from ^{27}Al MASNMR. Using this approach the structure of silicalite (closely related to ZSM-5) has been elucidated.[19] Prior to this study, it had been thought that Al was absent from the zeolite framework and instead present as Al_2O_3. The NMR data showed this to be incorrect; all the aluminium is present in the framework and is distributed over two distinct types of tetrahedral site. As well as being used for structural studies, MASNMR can be used to determine selectively the Si:Al ratio in the framework of the material[18] for Si:Al ratios in the range 1 to > 1000. This provides a useful alternative to X-ray fluorescence, which is not structurally selective. The future prospects for NMR investigations of zeolites look bright. The study of many of the exchangeable cations should be possible using this technique, and ^{13}C and ^{1}H NMR studies of catalytically important hydrocarbons within the zeolite are already proving rewarding. The incorporation of magnetic nuclei (e.g. Xe or N_2) as benign probes should provide further information on the nature of the environment within zeolites and important reactions such as dealuminization may be examined.

Although zeolite chemistry is the area of catalysis where NMR is playing a central role, the technique has also been exploited elsewhere. Stokes *et al.*[20] have used ^{195}Pt NMR to study the structure of supported Pt catalysts. In metals, a dominant effect on the chemical shift is the interaction of the nuclei with the conduction electrons via the s-orbital hyperfine coupling (Knight shift) which moves the resonance to higher fields. Atoms at the centre of the supported Pt particles experience the full Knight shift while those at the surface show a smaller effect. The peak profile is thus dependent on the size and geometry of the particles, as illustrated in Fig. 2.4, where NMR spectra of Pt catalysts with varying dispersions are presented. The chemical shifts exhibited span the range characteristic of the transition from isolated Pt species to bulk metal, thus the technique enables the types of Pt environment within the catalyst to be assessed. These authors also studied CO adsorption on the catalysts using double resonance techniques. They were able to show that although the NMR spectrum of Pt within the catalyst depends on the dispersion, the NMR spectra (and hence electronic structure) of the Pt involved in bonding the CO was independent of dispersion and very similar to that observed in large carbonyl species (e.g. $Pt_{38}CO_{44}^{2-}$).

NMR may also be applied as a complementary tool for the study of adsorption on catalysts. ^{1}H NMR has been used to study hydrogen adsorption on Rh/TiO$_2$

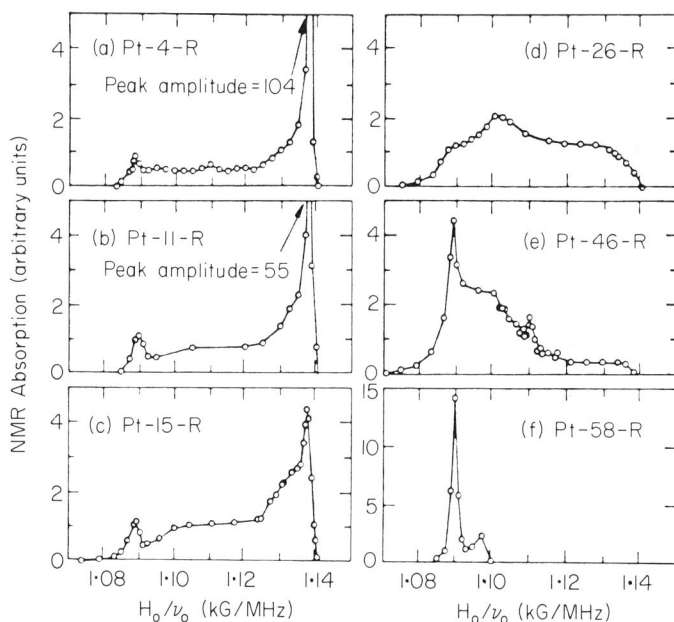

Fig. 2.4. NMR absorption line-shapes for six Pt/Al$_2$O$_3$ catalysts of increasing dispersions, (a)–(f). The resonance at 1.138 kG MHz^{-1} is characteristic of bulk Pt.[20] (Reprinted with permission from Elsevier Sequoia.)

catalysts.[21] It was shown that two types of absorbed hydrogen were present on the catalyst, one associated with the metal and the other reversibly bonded to the support. The presence of the metal was crucial to the formation of the second type of hydrogen, suggesting operation of a 'spill-over' mechanism. ^{13}C NMR provides a useful probe for the study of hydrocarbon adsorption. Dynamic investigations are possible, as shown by Nagy et al.,[22] who have used ^{13}C NMR to examine the isomerization of l-butene on mixed tin-antimony oxide catalysts. Other workers have identified the compounds formed within the adsorbed phase during the aldol condensation of acetone on alumina[23] and the oligomerization of olefins in ZSM-5 has also been studied by NMR.[24] The technique thus appears to offer considerable scope in the general area of catalyst characterization.

2.3   Neutron scattering

Centralized neutron scattering facilities are now available in the UK and neutron scattering offers a particularly promising technique for the *in situ* investigation of catalysts under live conditions. The basic experiment involves irradiation of the sample with a monochromatic thermal neutron beam and measurement of the angular and energy distribution of the scattered neutron intensity. The scattering field is divided into four experimental areas, depending on the energy and momentum transfer involved.

Inelastic scattering. Neutrons are scattered inelastically due to the excitation of vibrational modes within the sample, and hence a vibrational spectrum is obtained.

Quasi-elastic scattering. Energy broadening (< 4 meV) of the neutron beam is determined, providing information on the diffusive behaviour of adsorbed species.

Neutron diffraction. The familiar technique, which is complementary to X-ray diffraction, and particularly advantageous for the study of hydrogenous materials.

Small angle scattering. Diffusion by extended inhomogeneities in the sample characterizes pore and particle size distributions within catalysts.

It is thus apparent that a diverse range of information is available. Although other, perhaps more accessible techniques are obviously capable of imparting such information, neutron scattering can offer the experimentalist special advantages in particular circumstances. Most importantly, neutrons can penetrate the rugged containment vessels which are frequently used as catalytic reactors, providing unique opportunities for the study of working catalysts. Scattering cross-sections are such that the neutron can be used to highlight specific features of the sample. The enhanced sensitivity of neutron diffraction over X-ray methods for characterizing hydrogenous materials is established. This advantage also applies to the investigation of such materials, using neutron vibrational spectro-scopy. Useful cross-sectional differences between Si and Al exist, making neutron diffraction a preferable technique to X-ray diffraction for examining the structure of zeolites, where the question of Si/Al ordering is of continuing interest. With regard to inelastic scattering, there is an absence of selection rules and the quantitative prediction of spectral intensities is (relatively) straightforward. This is in contrast to optical spectroscopy where strong selection rules can prevent observation of bands of interest and the intensities depend on electronic factors not readily calculated. Neutron scattering methods are obviously at their most powerful when one or more of these advantages are fully exploited.

Neutron diffraction has proved particularly useful as a tool to supplement the information available from other techniques on the structure of zeolites.[25-29] Cheetham and workers[28] used the technique to characterize the structure of zeolite-A. The Si/Al ordering in this zeolite has been the subject of debate. One of the early studies using ^{29}Si MASNMR[30] indicated a 3:1 ordering scheme involving the presence of Al–O–Al bridges, in contradiction to Loewensteins rule, and this was supported by electron diffraction data.[31] However, Cheetham et al. showed that this conclusion was erroneous; the neutron results were hard to rationalize using any 3:1 scheme and suggested that there was strict alternation of Si and Al throughout the aluminosilicate framework. This conclusion is now generally accepted. The initial assignments of the NMR chemical shift ranges to particular geometric configurations were too restrictive, hence the NMR spectra were misinterpreted.

Most of the applications of neutron inelastic scattering (NIS) have involved the characterization of hydrogenous materials. The potential as an in situ probe is demonstrated by the investigation of hydrogen absorbed by MoS_2 at pressures up to 45 bar (comparable to industrial conditions for dehydrosulphurization processes). Variations in the vibrational spectrum of this material with increasing hydrogen pressure are shown in Fig. 2.5.[32] At low pressures, peaks at 622 and 872 cm^{-1} are observed, attributed respectively to S–H and Mo–H (or Mo–OH) deformations. As the pressure increases to 45 atm, a band at 400 cm^{-1} can be seen to gain intensity, as do MoS_2 lattice modes at c. 470 cm^{-1}. Clearly a further type of hydrogen is incorporated within the material. Without the ability to work at such high pressures, a species which could be important in the high pressure catalytic activity of MoS_2 would have been missed. A number of reports have recently appeared, showing the potential of NIS to characterize the nature of hydrocarbons chemisorbed on high surface area materials.[33-36] Using a finger-printing technique, it was demonstrated that low coverages of cyclohexane on Ni underwent decomposition to adsorbed benzene and hydrogen whereas, at higher coverages, reactions within the adsorbed phase formed cyclohexane. The ability to calculate NIS spectral intensities for particular models was exploited to determine the distortions (and hence loss of aromaticity) undergone when benzene adsorbs on Ni and Pt. The distortion on Pt was found to be the greater and this might relate to the more rapid hydrogenation rate of benzene on this metal. Kelley *et al.* have taken this type of work one stage further and have used NIS to characterize the adsorbed phase present on Ni in a H_2/CO mixture under reaction conditions.[37] Substantial quantities of adsorbed hydrocarbons were observed on the catalyst, when operating at 50% conversion.

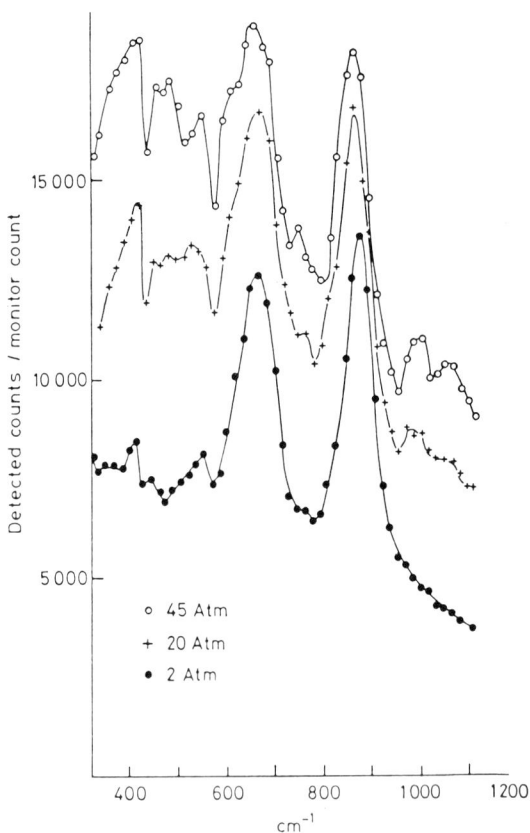

Fig. 2.5. NIS spectra of hydrogen absorbed by MoS_2 at varying pressures.[32] (Reprinted with permission from the Royal Society of Chemistry.)

The small angle scattering technique supplies similar information to that derived from the analogous X-ray method on the pore and particle size distributions within catalysts. Again, advantage can be taken of the cross-section differences which arise in the two approaches. From the point of view of catalysis, the interest in the future must lie in observing *in situ* changes in the physical structure of the catalyst during such processes as sintering and catalyst activation. The technique is also useful to characterize microporous materials, where the proven methods of isotherm analysis are less reliable and more tedious. In the case of mesoporous materials, the neutron small angle scattering data and isotherm analysis show good agreement, indicating the former technique is a reliable[38] method of determining surface areas. Having established this point, the same workers went on to characterize some microporous materials. They were able to show that neutron scattering provided a much more reliable method with which to examine such materials, since activated diffusion prevented filling of the micropores in the physisorption experiments. Finally, the advent of the quasi-elastic scattering technique to examine the diffusive behaviour of molecules within catalysts is to be welcomed. Diffusion processes are often invoked to explain various catalytic phenomena, e.g. apparent 'spill-over'. Unfortunately, although surface-diffusion of adsorbed species obviously represents an important step in catalytic action, it has hitherto been difficult to study directly under realistic conditions. Quasi-elastic scattering should therefore be able to cast light on this problem.

2.4   Electron microscopy

Electron microscopy plays an important role in catalyst characterization. In part this stems from the greater availability of the technique in some form to most workers, when compared to methods such as EXAFS which require highly specialized facilities. However, it is also the case that a modern electron microscope is a versatile instrument, ideally suited to the investigation of the poorly ordered materials (both on long-range and atomic scales) which constitute most heterogeneous catalysts. The physical structure of the catalyst may be straightforwardly determined. The atomic structure of supported catalyst particles down to 1 nm diameter can be obtained by utilizing direct imaging methods or electron diffraction analysis. X-ray emission and electron energy loss spectroscopies provide analytical information on the same scale. Bonding information similar to that provided by EXAFS and XANES is in principle available by analysing the energy loss structure at and above the various core-level edges. Using 'controlled atmosphere electron microscopy', processes occurring at temperatures up to 1000°C and pressures of 1 bar have been followed. This wealth of information supplied makes state-of-the-art electron microscopy a most useful technique with which to undertake structural studies in catalysis.

The principle of operation of the three distinct types of electron microscope is shown in Fig. 2.6. In the scanning electron microscope (SEM) electron optics act before the sample to produce a fine electron beam ($<$ 10 nm diameter, 5–50 keV) which is scanned over the surface of the sample by means of deflection coils. Electron scattering within the sample results in the production of back-scattered electrons, secondary electrons, emitted X-rays and optical photons. These are detected by suitable means and an image is displayed in a raster synchronous with

that of the probe scan. The resulting micrographs can resolve features on the
10–50 nm scale and are used primarily in catalysis for the examination of catalyst
topology. For example, SEM analysis of Rh/Pt gauzes used as ammonia oxidation
catalysts indicated that the deactivated form contained platelets and needles.[40]
X-ray microanalysis revealed that the latter consisted entirely of Rh_2O_3. Segregation of the two catalytic components was therefore indicated as being
responsible for the deactivation. Other typical applications include the
observation of sintering of large metal crystallites, and the catalytic growth of
carbon filaments.

While the SEM is a useful instrument, it is unable to provide information on an
atomic scale because of the limited resolution which is available. An ultimate
factor governing resolution is the beam spreading which occurs as the probe is
scattered within the sample. This problem is overcome by using extremely thin
samples (< 100 nm thick) and moving to a transmission mode, as shown in Fig.
2.6b, c. In the transmission electron microscope (TEM) a large stationary
illuminating electron beam is employed and lenses placed after the specimen are
used to produce a magnified image of the sample. Alternatively, diffraction
patterns can be recorded from selected areas of the sample. Various imaging
modes are employed to highlight differing aspects of the material under
investigation and resolution on a 2–3Å scale is possible, permitting the local
atomic structure to be examined. An alternative solution to the problem of high
resolution electron microscopy is provided by the scanning transmission electron
microscope (Fig. 2.6c) (STEM). This operates in a similar manner to the SEM,

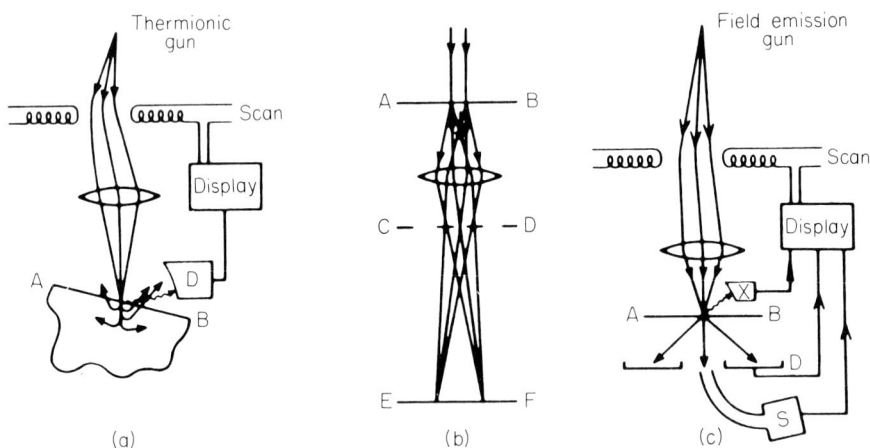

Fig. 2.6. Schematic diagrams showing the operation of (a) the scanning electron microscope (SEM),
(b) the conventional transmission electron microscope (TEM), (c) the scanning transmission
instrument (STEM).[39] AB, specimen plane; CD, diffraction plane; EF, image plane; D, electron
detector; S, electron spectrometer; X, X-ray detector. (Reproduced with permission from John Wiley
and Sons.)

except the electron probe is much finer (0.3 nm) and images are formed with the transmitted electron flux. Resolution is on an atomic scale as in the TEM and microanalysis is also naturally available, through the detection of emitted X-rays or energy loss electrons.

The ability to image directly even the smallest catalyst particles is now widely used to gain insight into their structure, particle size distribution and spatial disposition with respect to the support. A typical application of such work is provided by a recent paper[41] of Harris *et al.* who have studied the sintering of commercial Pt/Al$_2$O$_3$ catalysts. Using direct imaging methods, they obtained particle size distributions as a function of sintering time as shown in Fig. 2.7. From

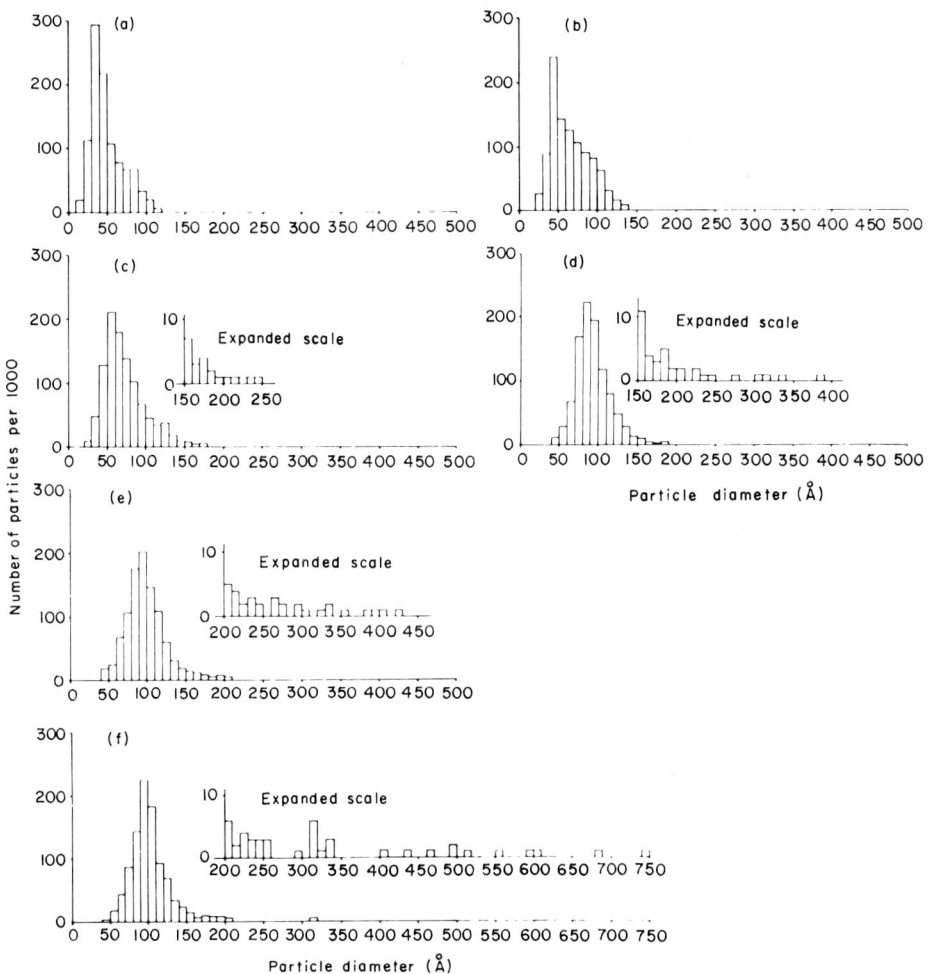

Fig. 2.7. Particle size distributions of Pt/Al$_2$O$_3$ catalysts for increased sintering times, (a)–(f), in air at 600°C. (Reproduced with permission from reference 41.)

the analysis of such data, the authors were able to conclude that the dominant mechanism initially involves particle migration but that this changes to inter-particle transport when the mean size exceeds 7.35 nm. The metal particles were found to be multiply twinned and electron microscopy shows this to be a common feature of supported catalysts.[42] This ability, to look at even the smallest atomic clusters, is demonstrated by the work of Schwank et al.[43] who observed the stability of triosmium clusters on Al_2O_3 supports using TEM.

The scope for direct imaging of lattice structures provides electron microscopy with considerable potential for the study of quasi-crystalline materials, i.e. most heterogeneous catalysts, which are not amenable for study by X-ray diffraction. High resolution electron micrographs of activated W–Ni hydrodesulphurization catalysts[44] have revealed the presence of WS_2 crystalline films, which were undetectable using X-ray methods. The pore structure in zeolite materials, which is responsible for the well known shape-selective catalytic properties, is readily viewed using direct imaging methods.[45] Crystalline regions of ZSM-5 in a sea of the amorphous material have also been observed,[45] providing an explanation as to why catalytic properties of the ordered material may be exhibited in situations where order is apparently absent. A particular feature of interest to catalytic chemists is the surface structure of the catalyst materials when in the supported state. Surface studies of bulk materials (section 3) indicate that interfaces may undergo a structural reorganization or 'reconstruction' under certain conditions and it is an intriguing question as to whether this carries through to supported materials. The surface structure of catalytic particles may be viewed by looking down the atom columns in the surface. Using this approach, it has been demon-strated that the (110) surfaces of small gold particles exhibit a '2×1' structure which is well known from surface studies of macroscopic gold specimens.[46] Although these observations were made for a highly idealized system, the work is of importance since it lends weight to the possibility that surface reconstructions may play an important role in the catalytic properties of small particles.

A particularly useful feature of the modern high resolution electron micro-scope is the ability to combine detailed structural information with micro-analytical results obtained by X-ray emission and electron energy loss spectroscopies. The high degree of spatial resolution offered has been exploited by Howie et al.[47] who have used the STEM technique to examine the distribution of Rh and Mn in bimetallic Rh:Mn catalysts, which are known to yield interesting product distributions in the Fischer–Tropsch synthesis process.[48] Typical X-ray spectra obtained for catalysts prepared in two different ways are shown in Fig. 2.8.[47] Catalysts prepared from the $RhCl_3$ precursor were found to contain an uncorrelated distribution of Rh and Mn; as is apparent from Fig. 2.8, the Mn is uniformly distributed within the Rh particles and the support. In contrast, the Mn is uniquely associated with the Rh, when a $Rh_4(CO)_{12}$ precursor is used. This

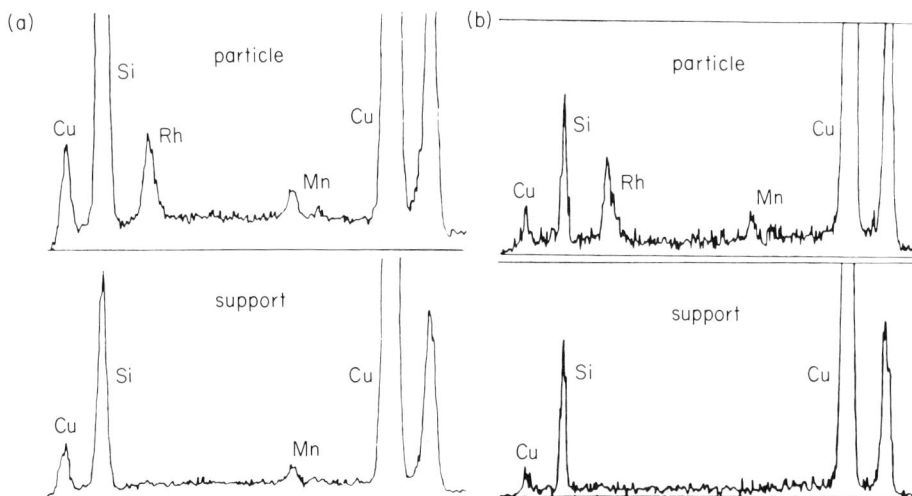

Fig. 2.8. X-ray spectra from RhMn/SiO$_2$ catalysts prepared (a) from RhCl$_3$ (b) from Rh$_4$CO$_{12}$. Spectra presented are for the catalyst particles and the SiO$_2$ support in the immediate vicinity. [48] (Reprinted with permission from Elsevier Sequoia.)

provides an explanation as to why the two catalysts display very differing selectivity patterns, despite containing the same atomic constituents.

Klier *et al.* have carried out detailed microanalytical studies of Cu/ZnO catalysts using STEM, to enquire into the synergic effect which the two catalytic constituents exert on each other with regard to their activity in the methanol synthesis process. [49] Real space images revealed that Cu and ZnO particles were in intimate contact within the catalyst and microanalysis indicated that substantial quantities of Cu were dissolved in the ZnO particles. The origin of the synergic effect was attributed to the presence of these dissolved Cu species although 'spill-over' of reactants and intermediates between the two phases present could also be responsible.

It seems likely that an increasingly important use of microanalytical facilities in the future will be the elucidation of the spatial distribution on an atomic scale of catalytic additives (e.g. promoters) which in many cases play an obscure but critical role in determining catalyst performance. Closely associated with the technique of electron energy loss spectroscopy for microanalysis is the use of the energy loss fine structure ('EXELFS') to extract structural information as is achieved in the EXAFS experiments. Although the full potential of the EXELFS approach has not yet been explored, the high degree of spatial resolution obviously offers an important advantage over most conventional EXAFS methods.

2.5   Other methods

In addition to the techniques described above, there are many other ways in which physicochemical information on catalysts and catalysis can be obtained. Perhaps the longest established and most widely used technique is transmission IR spectroscopy, which is available for characterization of the nature of adsorbed species which may form on high surface area catalytic materials. The advent of the computer and Fourier transform (FTIR) methods has done much to increase the power of IR spectroscopy in recent years. Much of the basic information acquired on the nature and configuration of adsorbed species of catalytic relevance, in which the general framework of catalysis is often discussed, has come from the application of the IR technique. By employing appropriate absorbates, information on the geometric and electronic structure of the catalyst itself may be obtained. Thus, for example, King & Peri[50] investigated NO adsorption on promoted Fe/Al_2O_3 catalysts and were able to demonstrate the co-existence of Fe^0 and Fe^{2+} adsorption sites and characterize the electron donor properties of the K promoter. The use of IR spectroscopy is not limited to static adsorption studies; the technique can also be employed as an *in situ* probe of catalytic activity under realistic working conditions. A classic example here is the reported study of the conversion of CO/H_2 mixtures to hydrocarbons over Ru and Fe catalysts using IR absorption spectroscopy.[51] The author was able to observe the growth of bands, arising from adsorbed CH_x intermediates involved in the Fischer–Tropsch synthesis process and thus cast light on the mechanism of the catalytic reaction.

Raman spectroscopy is also finding increased application in the field of catalysis and the development of multiplexing spectrometers now offers many advantages relating to improved signal–noise ratios and data acquisition rates. In marked contrast to IR, the Raman spectrum of water is weak and so studies of surface species at the aqueous/solid interface are possible, especially when use is made of the so-called 'enhanced Raman scattering' phenomenon. This obviously offers possibilities for *in situ* studies of electrocatalysis and wet catalyst preparation procedures. Because of differences in the intensity of the background absorption, Raman methods can be more useful than IR techniques for the study of supported oxide catalysts but may be problematic when applied to their metal counterparts. This is reflected in the areas in which Raman spectroscopy has made the greatest contributions to date. A number of Raman investigations of zeolites and incorporated molecules has been reported, which cast light on zeolite structure and the nature of the environment within the cages (e.g. references 52, 53). The *in situ* capabilities of the technique were shown by Grasselli[54a] who studied phase transitions in $BiMoO_4$ catalysts in this way and by Schrader and Cheng[54b] who examined the sulphiding of $Co\text{-}Mo/Al_2O_3$ catalysts. These authors were able to follow the various stages in the sulphidization process of the individual phases which constitute the oxidic form of the catalyst and infer that the presence of Co inhibits the growth of large crystallites of MoS_2.

Mossbauer spectroscopy is another viable technique when Mossbauer-active elements are present within the catalyst. The classic example of the application is the study of Fe catalysts, but many other elements are amenable for investigations; even when 'Mossbauer' elements are not present, they may be incorporated in trace amounts as benign probes, this representing a useful extension of the technique. By measurement of parameters such as isomer shift, quadrupole and magnetic hyperfine splittings, the chemical state and the nature of the environment of the element in focus may be ascertained. Measurement of magnetic hyperfine splittings can provide a useful means for ascertaining particle sizes. Finally, *in situ* studies are clearly possible in view of the penetrating power of the gamma ray probe employed. Mossbauer techniques have to date proved useful in the characterization of the structure and chemistry of mixed oxide selective oxidation catalysts,[55] Fe-based bimetallic catalysts,[56] zeolites,[57] and supported Fe catalysts.[58] Sinfelt[59] demonstrated how Fe could be incorporated into Pt–Ir reforming catalysts to obtain information on the degree of aggregation of the two active metal components.

All of the physical techniques described so far are essentially methods for characterizing the bulk properties of materials; they find application in catalyst characterization (where fundamentally it is the surface properties which are of interest) because the highly dispersed nature of such materials removes the distinction between 'bulk' and 'surface' regions. As described in the latter part of this article, many powerful techniques have been developed which possess intrinsic surface sensitivity, even when applied to the study of low surface area microscopic samples. Such methods include secondary ion and fast atom bombardment mass spectrometries (SIMS, FABMS), scanning Auger microscopy (SAM) and X-ray photoelectron spectroscopy (XPS, 'ESCA'). These techniques may also be used for post-mortem analyses of dispersed catalysts, from which the chemical constitution of the external surfaces of the catalyst sample can be obtained. The best established is ESCA, which can straightforwardly determine chemical composition on a sub-monolayer level and measure ion charge states (although problems in spectral interpretation can arise). An example of the application of this technique comes from the work of DeVries *et al.*[60] on PtO_2–MoO_3/Al_2O_3 catalysts. They studied the low temperature reduction of this catalyst with varying Pt:Mo ratios and recorded the spectra illustrated in Fig. 2.9. As the Pt loading increases, the binding energy of the Mo(3d) levels is reduced. This was interpreted as being due to an increase in the concentration of Mo(IV) and Mo(V) species present within the catalyst at the higher Pt loading. Ertl and co-workers[61] have combined the techniques of ESCA, SAM and SEM in a study of triply promoted ammonia synthesis catalysts. Using such methods they were able to obtain both the chemical states and the spatial distribution of the catalyst constituents (Fe, K, Ca, Al, O) within the surface layers. Their conclusions were

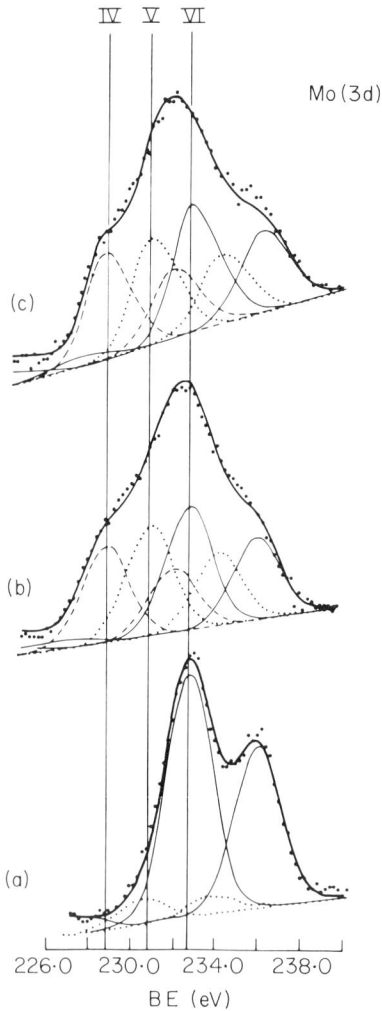

Fig. 2.9. Mo(3d) XPS spectra of partially reduced PtO_2–MoO_3/Al_2O_3 catalysts with increasing Pt loadings, (a)–(c).[60] (Reprinted with permission from Academic Press.)

found to be in full qualitative accordance with previous reports, which were based on rather indirect methods (selective adsorption, etc.).

3   Studies on model systems

3.1   *Experimental approach*

Although the application of physical techniques, such as have been described above, has made a major contribution to our understanding of the structure and reactivity of dispersed catalyst systems, there are great difficulties associated with the study of these high surface area materials. Their complete structural characterization is still impossible because of the lack of uniformity and definition on an atomic scale. An approach is clearly required in which the structure and composition of the catalyst surface is well-defined and the three basic steps associated with catalytic action, viz adsorption of reactants, interconversion of surface intermediates, desorption of products, can be completely characterized. The link between (electronic and geometric) structure, composition and surface reactivity may then be established and the important parameters governing catalyst performance understood. This has become known as the 'surface science' approach to catalysis. The experimental method is to work with bulk small-area (10^{-4} m^2) single crystal samples. The surfaces of these specimens are structurally uniform and may be chosen to expose any particular crystallographic plane of the material—control of geometry is thus achieved. The atomic integrity of the surface may be maintained for extended periods by working in ultra-high vacuum (UHV) ($< 10^{-9}$ mbar) conditions and Ar$^+$ etching procedures are used to remove unwanted surface contaminants. The surface composition can be modified by controlled deposition of additives (e.g. catalyst promoters) from the vapour phase. Model catalyst surfaces of any desired structure and composition can thus be prepared. Using conventional UHV chambers the reactive properties of these surfaces in low pressure reactive gaseous mixtures (10^{-9}–10^{-3} mbar) can then be characterized. To carry out such work, experimental techniques are required which are capable of examining the structure, composition and adsorption–desorption properties of the surface phases involved. The techniques described in section 2 are generally inapplicable since they either do not possess sufficient sensitivity or the surface information is hopelessly buried in the signal emerging from the bulk of the sample. Nevertheless, surface scientists have been uniquely successful in developing an almost bewildering number of suitable analytical probes, which are capable of supplying the required information in enormous detail. These techniques generally involve irradiation of the sample with electron, ion, atom or photon beams of appropriate energy and detection of some signal, which results from the interaction of the radiation with the surface layers. These methods are described in detail elsewhere,[62–64] but a selection of the most useful is listed in Table 2.1.

The major problem associated with the use of these surface techniques is the short mean free path of the probe particles employed, at high gas pressures; this effectively limits the pressure which can be used in the experiments to 10^{-3} mbar at most and often very much less. Unfortunately, most industrial catalytic processes of practical interest do not occur to a detectable extent under these low pressure conditions. The solution to this problem is to use a high pressure environmental cell, which is directly attached to the UHV surface analysis

Table 2.1. Surface techniques

Surface analysis technique	Physical basis	Information supplied
Low energy electron diffraction (LEED)	Elastic back-scattering of low energy electrons.	Atomic surface structure of surfaces and adsorbates.
Surface EXAFS (SEXAFS)	EXAFS, using electron or ion currents to determine surface absorption cross-sections.	Similar to LEED.
Auger electron spectroscopy (AES)	Low energy electron emission from surface atoms excited by electron or photon.	Surface composition.
Secondary ion mass spectrometry (SIMS)	Ion-beam-induced ejection of ion clusters from surface.	Surface composition and state of aggregation of substrates and adsorbates.
X-ray and ultraviolet photoelectron spectroscopies (XPS, UPS)	Photoemission of electrons from surface atoms.	Surface composition and structure of substrates and adsorbates. Electronic structure of surface species.
High resolution electron energy loss spectroscopy (HREELS)	Vibrational excitation through inelastic scattering of low energy electrons at surface.	Structure and bonding of adsorbates.
Thermal desorption spectrometry (TDS)	Thermally induced desorption and decomposition of adsorbed species.	Kinetics and energetics of adsorption–desorption processes and surface reactions.

chamber by means of a suitable isolation valve.[65] Samples can be scrutinized using the techniques of surface science and then transferred to the high pressure cell, where catalytic reactions are carried out. Catalytic activity and selectivity are assessed by periodic sampling of the cell contents using a gas chromatograph and post mortem analysis of the surface is performed, after transferring the crystal back to the low pressure chamber. In this way catalytic performance may be correlated with measured variations in surface geometric and electronic structure, composition and basic reactive properties. Reaction rates, activation energies and product distributions which have been measured for model systems have been found to be identical to those observed for dispersed catalysts in a number of instances. This confirms that single crystal substrates are capable of providing appropriate models for the investigation of processes occurring on large-area practical catalysts. The surface science approach has made a major impact in recent years on our fundamental understanding of catalytic processes. It is hoped to demonstrate this point in the remainder of this article by considering a limited number of areas of the work where important contributions have been made.

3.2   The geometric factor in catalysis

Early theories in catalysis attempted to attribute the difference in catalytic properties of various materials to a dependence of reaction rate on either the geometric structure of the surface ('geometric factor'), or the electronic properties ('electronic factor'). Such a formulation is of limited value since surface geometry cannot be varied independently of localized electronic structure. Nevertheless, the occurrence of 'geometric effects'—a dependence of reaction rates on surface structures as well as composition—is of continuing interest to the catalytic chemist. The existence of a geometric effect may be demonstrated indirectly from studies of real catalytic systems through a variation of the specific catalytic rate with particle size. Using such a criterion, catalytic reactions are classified as being 'demanding' or 'facile'. However, the lack of structural definition in dispersed systems makes it impossible to study geometric factors in any detail. In contrast, the 'surface science' approach is ideally suited to such work, since surface crystallography may be accurately controlled.

A simple example of a geometric effect concerns the dissociation of CO, which is believed to be a crucial step in the Fischer–Tropsch process for the synthesis of hydrocarbons. Surface science studies have convincingly demonstrated that CO dissociation on transition metals at the left of the periodic table is rapid, whereas it probably does not occur on elements situated to the right of a dividing line defined by Co, Ru, Os. Co is one of the borderline elements and CO adsorption has been studied on the smooth Co(0001) and stepped Co(10$\bar{1}$2) surfaces.[66,67] While dissociation of CO on the basal plane was undetectable, dissociation probabilities of ~0.1 were observed during thermal desorption of CO from the stepped surface. Clearly, a 'geometric factor' is displayed; although its demonstration is straightforward, it is a much more difficult matter to predict the role such an effect might play in the catalytic chemistry of Co. This example therefore serves to demonstrate both the power of surface science studies in catalysis, and some of their limitations. Dissociative nitrogen chemisorption on Fe is another reaction which has been demonstrated to be highly structure sensitive.[68] The rate of chemisorption on the (111), (100) and (110) planes of Fe was found to vary in the approximate ratio 60:3:1. In this particular instance, the high pressure ammonia synthesis activity of the three single crystal Fe planes has also been measured using apparatus similar to that described in section 3.1, and a direct correlation was found between these sticking probabilities and the overall catalytic rates.[69] This shows the usefulness of such high pressure measurements in elucidating the catalytic significance of surface chemical effects deduced at much lower pressures. Ertl et al.[70] have investigated in detail the elementary steps involved in the dissociative chemisorption of nitrogen. The picture to emerge is that the first stage involves the rapid physisorption of nitrogen molecules into a weakly bound state on the Fe surface. Transfer to a strongly bound, dissociated chemisorbed state then occurs subsequently. This second step is an activated process and the variation in activation energy with crystal plane accounts for the overall structure sensitivity of the sticking probability for chemisorption. A potential energy diagram for the reaction, based largely on thermal desorption studies, is illustrated in Fig. 2.10. The same authors also studied the role played by potassium, an important catalyst promoter for ammonia synthesis, in the adsorption process. Its effect is to lower the activation energy to dissociation by increasing the depth of the physisorption well (Fig. 2.10). Hence, the nitrogen chemisorption probability

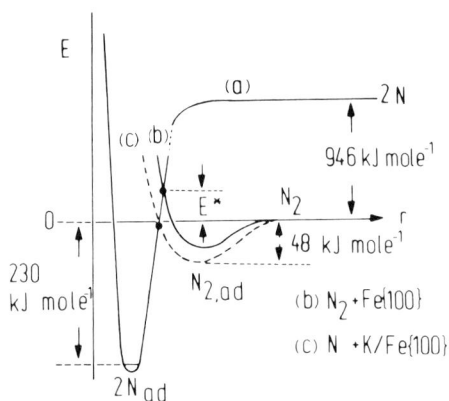

Fig. 2.10. Potential energy diagram for the adsorption of nitrogen on Fe(100) (b) in the absence (c) in the presence of K(70). (Reproduced with permission from the North Holland Publishing Co.)

increases. Interestingly, the magnitude of this promotional effect was also found to be structure sensitive, being largest on the most inactive (110) plane. The structure sensitivity of the ammonia synthesis reaction is thus largely reduced when the promoter is incorporated into the catalyst.

Other examples of the structure sensitivity of surface chemical processes observed using single crystals are described in section 3.4 of this chapter. Clearly geometric effects in catalysis are important and their study using the techniques of modern surface science represents a most powerful approach.

3.3   Catalyst promoters and poisons
Working catalysts often contain many foreign species (incorporated intentionally or otherwise) which play an important role in determining catalyst performance. Modern surface science methods have begun to provide a good understanding of the molecular basis of the effects which such species bring about. Chemical promoters and poisons are often highly electropositive (-negative) species and early theories of catalysis suggested they functioned by feeding electrons into or

withdrawing electrons from the delocalized electron bands in solids. The picture which is currently emerging suggests this to be incorrect; the important electronic effect is to modify the *localized* electronic properties around the particular adatoms. The straightforward blocking of specific binding or reaction sites is also important.

Most of the surface analytical studies carried out to date on catalyst poisoning have looked at the surface effects of sulphur and halogen species. The first stage is to characterize the surface chemistry and adsorption energies of these species and this has been achieved in many instances (reference 71 contains a detailed review of the sulphur work and the reader is referred to an article by Reed *et al.*[72] for an insight into the halogen studies). A number of points have emerged of catalytic significance. The adsorption of these species is not limited to particular geometric sites on surfaces; instead strongly bound close-placked adsorbed layers form with a geometry mainly defined by the size of adatoms. Complete catalyst poisoning is therefore possible through simple geometric blocking. The energetics of formation of thin film sulphide and halide phases on metals is rather similar to that associated with formation of the bulk compounds. The presence of such phases under catalytic conditions may thus be predicted using bulk thermodynamics. Restructuring of the substrate surface may be brought about by the presence of these species, pointing to the possible occurrence of a geometrical factor and the morphology of supported metal crystallites may be affected.[73] With regard to co-adsorption of the species involved in catalytic reactions, the principal effect is one of site-blocking, which serves to break up the 'ensembles' of active sites required for particular surface reactions. For example, Foord & Lambert demonstrated that oxygen dissociation on chromium oxide is inhibited in the presence of chlorine, because of the requirement for adjacent active sites for the dissociation step[74] and Madix *et al.* have reported similar effects for methoxy intermediates adsorbed on Ni(100).[75]

The majority of work in surface science on catalyst promoters has concentrated on the action of alkali metals. The effect of K on the dissociative sticking probability of nitrogen has already been described above and a similar effect has also been observed for dissociative CO adsorption.[76] This has obvious possible consequences for the Fischer–Tropsch process. A number of studies on the promotional effects of alkali species on Ag, with regard to the ethylene oxide synthesis process, has been reported. Alkali metals appear to both stabilize dioxygen species present on the Ag surface and promote formation of subsurface oxygen.[77,78] The exact significance of these effects will become clearer when the reaction mechanism for the synthesis of ethylene oxide is better understood. Foord & Lambert[79] have studied the surface chemical effects induced by sodium adsorbed on Cr_2O_3. They found that the alkali species served to stabilize formation of Cr (VI) centres at the oxide surface. This has importance since the concept of variable valency is central to many theories concerning the catalytic properties of mixed transition metal oxides. It is thus apparent that the presence of alkali species may have a number of wide-ranging effects in catalysis.

3.4   Bimetallic systems

Bimetallic catalysts are of interest since they can offer advantages in terms of improved selectivity, activity and stability, over their single metal counterparts.[80] The attention of surface scientists has recently turned to the study of such systems in order to investigate the theories proposed to explain their catalytic properties. The need to minimize surface free energy causes the surface alloy composition to differ from the bulk constitution, and surface segregation models therefore form an implicit part of the theory of bimetallic catalysis. Various theories have been proposed to explain the modifications in catalytic properties which alloy formation brings about. One of the first theories involved the rigid-band approximation which supposes the band structure does not alter on alloying; alloying then merely affects the band occupancy (and hence the electron donor–acceptor properties of the material). Later theories supposed that atoms retain most of their individuality in alloys, and that the reactivity changes are caused by modifications in the geometric distribution of the particular atoms within the alloy ('ensemble effect') or localized electron transfer ('ligand effect') between adjacent atoms of differing electronegativity. Surface science methods have provided the major tools with which to investigate the validity of these theories.

Surface segregation is readily studied experimentally on macroscopic samples, using such surface analytical techniques as AES, XPS and ion scattering spectroscopy (ISS) which can be combined to produce a concentration profile through the outermost layers of the material. One of the simplest theoretical models to explain segregation is the 'regular solution theory' which makes predictions based on the heat of mixing of the alloy components and their relative surface tensions.[81] The other major approach is the lattice strain theory[82] which recognizes that the lattice misfit introduced by solutes will tend to favour their segregation in order to relieve strain. Surface analytical studies of segregation effects have been carried out for a variety of alloy systems such as Cu–Ni, Au–Ag, Au–Ni, Cu–Au, Fe–Sn and Pt–Sn.[83] The results presently suggest that the two theories recognize the basic driving forces behind surface segregation and are qualitatively correct when used in combination.

The application of physical techniques has also helped to explain the advantageous chemical properties of alloy catalysts. Classic photoemission studies[84] revealed that the rigid band models were incorrect. Schematic photoemission spectra are illustrated in Fig. 2.11; apart from a second order effect associated with a change in the d-band width, the spectrum of the alloy is essentially the sum of the spectra of the two separate alloy components, in

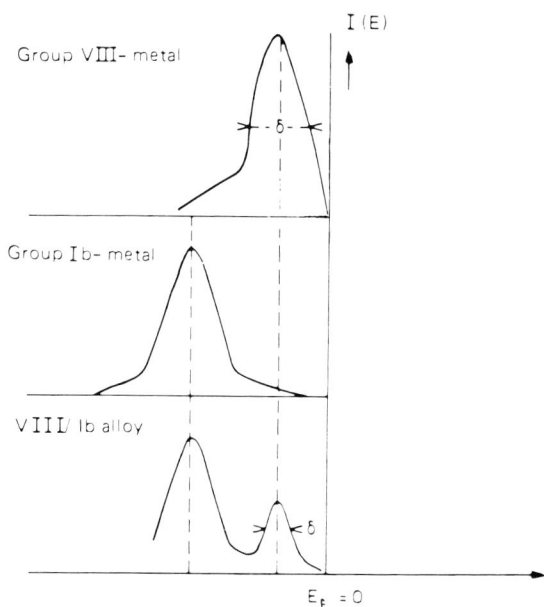

Fig. 2.11. PES spectra of Group VIII and Group Ib metals and their alloys.[85] (Reprinted with permission from Academic Press.)

contradiction to the rigid band theory. Even a pronounced ligand effect is in doubt in the case of the endothermic or moderately exothermic alloys of interest in catalysis. Thus, Ponec *et al.* observed no shift in the IR stretching frequency of CO adsorbed on Pt when Cu was incorporated, provided account was taken of dipole–dipole coupling effects.[86] This indicates that the electron donor properties of Pt were unaffected by alloying with Cu. Small changes in heats of adsorption upon alloy formation have been noted which purport to show the existence of a ligand effect. However, it is at present unclear to what extent these should be alternatively attributed to a change in the lateral interactions within the adsorbed layer or the nature of the bonding site. By far the most pronounced effect of alloy formation appears to be the induced variation in the ensemble size of the active atoms in the alloy. Alloying obviously breaks up the larger atomic ensembles of the individual constituents and thus selectively inhibits structure sensitive catalytic processes which depend upon their presence. This effect is particularly well demonstrated by the work of Ertl *et al.*[87] who have characterized the surface chemistry of Ru–Cu interfaces in great detail.

3.5   Catalysis on single crystals
Most of the fundamental surface science studies related to catalysis have been involved with the characterization of the nature and properties of metal catalyst

surfaces and adsorbed species. While such work is invaluable, its usefulness is greatly increased if the impact of the observations on actual catalytic processes can be assessed. This is best carried out by measuring catalytic rates on the actual single crystal surfaces chosen for initial surface characterization, using the high pressure–low pressure transfer technology described in section 3.1. To date, this has only been carried out in a limited number of instances, although the approach is becoming increasingly popular.

The major work here has undoubtedly been that of Somorjai and co-workers[88,89] who have carried out detailed studies of hydrocarbon conversion over Pt. The particular aim has been to correlate catalytic properties with the structure of the Pt surfaces, in particular the 'terrace' and 'kink' site concentrations. Typical results are shown in Fig. 2.12 for the aromatization of n-hexane, whence it can be seen that the reaction displays considerable structure sensitivity. Using this type of data, it was shown that surface steps are active for the bond making/breaking of C–C and C–H entities, while kink sites have a unique activity for C–C bond cleavage. These authors have observed a pressure and temperature dependence in the structure sensitivity of hydrocarbon conversions, which they associate with the presence of a carbonaceous overlayer, seen during post mortem AES analysis of the Pt crystals. The effect of this layer is to smooth out surface irregularities and thus reduce the sensitivity of the reaction to the underlying Pt structure.

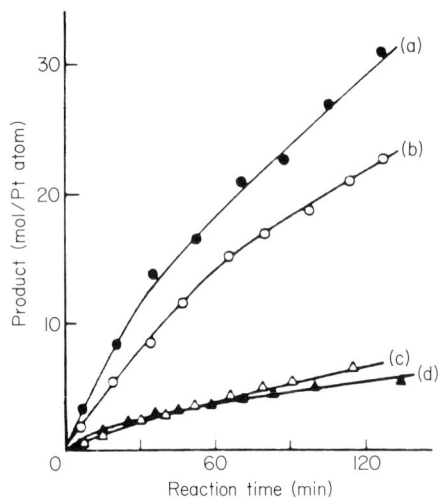

Fig. 2.12. Plot of the accumulation of product with time for the aromatization of n-hexane over (a) Pt (10, 8, 7); (b) Pt(111); (c) Pt(100); (d) Pt(13, 1, 1). (Reprinted from reference 90 with permission from the Institute of Physics.)

Goodman *et al.* have studied methane formation from CO/H_2 mixtures over Ni single crystals and have used AES to determine the steady state composition of the catalyst surface.[91] The kinetic data obtained are shown in Fig. 2.13; remarkably good agreement between the single crystal data and that from high surface area catalysts is observed demonstrating the structure insensitivity of the reaction. Their AES data revealed that the Ni surface was covered in 5% monolayers of carbide under steady state conditions. This carbide layer was hydrogenerated at the same rate at which methane is formed, suggesting strongly that it is the vital intermediate in the overall catalytic tranformation.

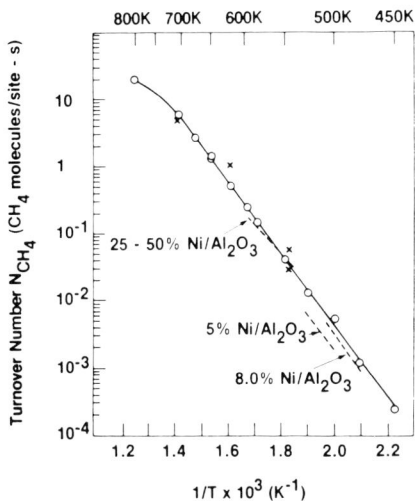

Fig. 2.13. Arrhenius plot for CH$_4$ synthesis over (o) Ni(100), (x) Ni(111), and three high surface area Ni catalysts. (Reprinted from reference 92 with permission from Elsevier.)

Single crystal studies of the silver-catalysed epoxidation of ethylene have made a major contribution to elucidating the reaction mechanism.[93] Disagreement has long existed in the literature as to the nature of the adsorbed oxygen species responsible for the selective oxidation of ethylene, the debate centering on whether the desired species is atomic or diatomic. Although the subject of many mechanistic studies on supported silver catalysts, no conclusive picture emerged. In the surface science work it was possible to prepare oxidized silver surfaces on which mixtures of atomic and diatomic oxygen species were present; alternatively the atomic species alone could be adsorbed. It was found that the same epoxidation activity was exhibited, irrespective of the presence of diatomic oxygen. This indicates that the presence of atomic oxygen is sufficient for the selective oxidation of ethylene and, indeed under the conditions of their experiments, the molecularly adsorbed oxygen played no part in the selectively catalysed process.

Ertl *et al.*[94] have studied the oxidation of CO on Pt(100) and provided a most intriguing insight into the connection between surface structure and catalytic activity. The unusual feature of this particular catalytic process is the fact that it exhibits pronounced, sustained oscillations in the reaction rate, under certain conditions. This may be understood theoretically as arising from a strong non-linear dependence of some surface process on coverage of the reactants. Pt(100) is well known to undergo a surface reconstruction and LEED revealed that phase transformation between the reconstructed and non-reconstructed forms accompanied the oscillation in reaction rate. This arises because the stability of a particular surface configuration depends on the nature of the adsorbed species present. Since the rate of individual processes involved in the overall catalytic reaction depends upon the structure of the Pt, the strong non-linear variations required for sustained oscillations in rate are explained.

4   Concluding remarks

The development of new physical techniques, such as those based on neutron scattering and intense X-ray sources, and the extension of traditional methods for the study of dispersed catalysts represent considerable advances in the field of catalyst characterization. A discernible move is taking place towards *in situ* studies: this is particularly welcome. Although many of the newer physical techniques for examining catalysts are presently of interest to those concerned with 'fundamentals' they can offer real use in the industrial practice of catalysis if only through their applications as 'finger-printing' techniques. Clearly under such circumstances, availability and speed of data analysis are important criteria to be considered when assessing the possible use of a particular method.

Simultaneous to this development of techniques for studying real catalysts has been an increase in the level of sophistication in the surface science approach to catalysis. The use of high-pressure environmental cells has done much to solve the so-called 'pressure-gap' problem and the topics of study are becoming ever more relevant. While surface science still does not appear to represent a practical main-stream approach to the development of new catalysts the information thrown up has a major impact on the understanding of catalytic phenomena. The increasing application of physical techniques for the characterization of catalysts should provide an added impetus to the study of model systems; as the knowledge of catalyst structure increases, so will the realism in the models chosen for investigation in surface science.

5   References

1   Bowker, M. *Vacuum* 1983, **33,** 669.
2   Satterfield, S.S. *Heterogeneous Catalysis in Practice.* McGraw-Hill, New York, 1980.
3   Teo, B.K. & Joy, B.C. (Ed.) *EXAFS Spectroscopy, Techniques and Applications.* Plenum, New York, 1980.
4   Winick, H. & Doniach, S. (Ed.) *Synchrotron Radiation Research.* Plenum, New York, 1980.
5   Sinfelt, J.H., Via, G.H. & Lytle, F.W. *J. Chem. Phys.* 1982, **76,** 2779.
6   Joyner, R.W. & Meehan, P. *Vacuum* 1983, **33,** 691.

7   Werfelmeier, W. *Z. Physik.* 1937, **107**, 332.
8   Sinfelt, J.H., Via, G.H. & Lytle, F.W. *J. Chem. Phys.* 1980, **72**, 4832.
9   Sinfelt, J.H., Via, G.H., Lytle, F.W. & Greegor, R.B. *J. Chem. Phys.* 1981, **75**, 5527.
10 Parham, P.G. & Merill, R.P. *J. Catal.* 1984, **85**, 295.
11 Kozlowski, R., Pettifer, R.F. & Thomas, J.M. *J. Phys. Chem.* 1983, **87**, 5172.
12 Vant Blik, H.F.J., Van Zon, J.B., Huizinga, T., Vis, J.C., Koningsberger, D.C. & Prins, R. *J. Phys. Chem.* 1983, **87**, 2264.
13 Fleming, I. & Williams, D.H. *Spectroscopic Methods in Organic Chemistry.* McGraw-Hill, London, 1973.
14 Andrew, E.R. *Arch. Sci. (Geneva),* 1959, **12**, 103.
15 Lowe, I.J. *Phys. Rev. Lett.* 1959, **2**, 285.
16 Duncan, T.M. & Dybowski, C. *Surface Sci. Rep.* 1983, **3**, 1.
17 Engelhardt, G., Kunath, D., Samoson, A., Tarmak, M., Magi, M. & Lippmaa, E. *Workshop on Adsorption of Hydrocarbons in Zeolites.* Berlin-Adlershof, Nov. 19–22, 1979.
18 Fyfe, C.A., Thomas, J.M., Klinowski, J. & Gobbi, G.C. *Angew. Chem. Int. Ed. Eng.* 1983, **22**, 259.
19 Fyfe, C.A., Thomas, J.M., Gobbi, G.C., Klinowski, J. & Ramdas, S. *Nature (London)* 1982, **296**, 530.
20 Stokes, H.T., Makowka, C.D., Wang, Po-Kang, Rudaz, S.L., Slickter, C.P. & Sinfelt, J.H. *J. Mol. Catal.* 1983, **20**, 321.
21 Apple, T.M., Gajando, P. & Dybowski, C. *J. Catal.* 1981, **68**, 103.
22 Nagy, J.B., Abov-Kais, A., Guelton, M., Harmel, J. & Derouane, E.G. *J. Catal.* 1982, **73**, 1.
23 Bell, V.A. & Gold, H.S. *J. Catal.* 1983, **79**, 286.
24 van den Beng, J.R., Woltzhoizen, J.P., Clague, A.D.H., Hays, G.R., Huis, R. & van Hooff, J.H.C. *J. Catal.* 1983, **80**, 130.
25 Kvick, A. & Smith, J.V. *J. Chem. Phys.* 1983, **79**, 2356.
26 Bennett, J.M., Blackwell, C.S. & Cox, D.E. *ACS Symp. Ser.* 1983, **218**, 143.
27 Cheetham, A.K. & Eddy, M.M. *ACS Symp. Ser.* 1983, **218**, 131.
28 Cheetham, A.K., Eddy, M.M., Jefferson, D.A. & Thomas, J.M. *Nature (London)* 1983, **299**, 24.
29 Adams, J.M., Hasledon, D.A. & Hewatt, A.W. *J. Solid State Chem.* 1982, **44**, 245.
30 Engelhardt, G., Zeigen, D., Lippmaa, E. & Magi, M. *Z. Anorg. Allg. Chem.* 1980, **468**, 35.
31 Bursill, L.A., Lodge, E.A. & Thomas, J.M. *J. Phys. Chem.* 1981, **85**, 2409.
32 Vasudevan, S., Thomas, J.M., Wright, C.J. & Sampson, C. *J. Chem. Soc. Chem. Commun.* 1982, 419.
33 Howard, J., Robson, K., Waddington, T.C. & Kadir, Z.A. *Zeolites.* 1982, **2**, 2.
34 Howard, J., Robson, K. & Waddington, T.C. *Zeolites.* 1981, **1**, 175.
35 Candy, J.P., Jobic, H. & Renouprez, A.J. *J. Phys. Chem.* 1983, **87**, 1227.
36 Jobic, H. & Renouprez, A. *Surface Sci.* 1981, 53.
37 Kelley, R.D., Cavanagh, R.R. & Rush, J.J. *J. Catal.* 1983, **83**, 464.
38 Wright, C.J. In *Characterisation of Catalysts,* Thomas, J.M. & Lambert, R.M. (Eds), p. 179. Wiley, Chichester, 1980.
39 Howie, A. In *Characterisation of Catalysts,* Thomas, J.M. & Lambert, R.M. (Eds), p. 90. Wiley, Chichester, 1980.
40 Acres, G.J.K., Bird, A.J., Jenkins, J.W. & King, F. In *Characterisation of Catalysts,* Thomas, J.M. & Lambert, R.M. (Eds), p. 72. Wiley, Chichester, 1980.
41 Harris, P.J.F., Boyes, E.D. & Cairns, J.A. *J. Catal.* 1983, **82**, 127.
42 Howie, A. In *Characterisation of Catalysts,* Thomas, J.M. & Lambert, R.M. (Eds), p. 98. Wiley, Chichester, 1980.
43 Schwank, J., Allard, L.F., Deeba, M. & Gates, B.C. *J. Catal.* 1983, **84**, 27.
44 Thomas, J.M., Millward, G.R. & Bursill, L.A. *Phil. Trans. R. Soc. (London)* A300 1981, 43.
45 Thomas, J.M., Millward, G.R., Ramdas, S., Bursill, L.A. & Audier, M. *Faraday Disc. Chem. Soc.* 1981, **72**, 345.

46 Marks, L.D. & Smith, D.J. *Nature* 1983, **303**, 316.
47 Pennycock, S.P., Howie, A., Shannon, M.D. & Whyman, R. *J. Mol. Catal.* 1983, **20**, 345.
48 Wilson, T.P., Kasai, P.H. & Ellgen, P.C. *J. Catal.* 1981, **69**, 193.
49 Dominique, J.M., Simmons, G.W. & Klier, K. *J. Mol. Catal.* 1983, **20**, 369.
50 King, D.L. & Peri, J.B. *J. Catal.* 1983, **79**, 164.
51 King, D.L. *J. Catal.* 1979, **56**, 287.
52 Roozeboom, F., Robson, H.E. & Chan, S.S. *Zeolites.* 1983, **3**, 321.
53 Pechar, F. & Rykl, D. *Zeolites.* 1983, **3**, 329.
54 (a) Grasselli, J.G., Hazle, M.A.S., Mooney, J.R. & Mehici, M. *Sohio Rep.* 5705, Aug. 1979.
54 (b) Schrader, G.L. & Cheng, C.P. *J. Catal.* 1984, **85**, 488.
55 Matsuwa, I. *Proc. 6th. Int. Congress on Catalysts,* **2**, p. 189. Chemical Society, London, 1977.
56 Garten, R.L. *Mossbauer Effect Methodology,* 1976, **10**, 69.
57 Bodant, P., Nagy, J.B., Debras, G., Gabelica, Z., Derouane, E.G. & Jacobs, P.A. *Bull. Soc. Chim. Bldg.* 1983, **92**, 711.
58 Boudart, M., Dellouille, A., Dumesic, J.A., Khammouma, S. & Topsoe, H. *J. Catal.* 1975, **37**, 486.
59 Garten, R.L. & Sinfelt, J.H. *J. Catal.* 1980, **62**, 127.
60 DeVries, J.E., Yao, H.C., Baird, R.J. & Gandhi, H.S. *J. Catal.* 1983, **84**, 8.
61 Ertl, G., Prigge, D., Schloegl, R. & Weiss, M. *J. Catal.* 1983, **79**, 359.
62 Ertl, G. & Kuppers, J. *Low Energy Electrons and Surface Chemistry.* Verlag Chemi, Weinheim, 1974.
63 Roberts, M.W. & McKee, C.S. *Chemistry of the Metal Gas Interface.* Clarendon Press, Oxford, 1978.
64 Kane, P.F. & Larabee, G.B. *Characterisation of Solid Surfaces.* Plenum, London, 1974.
65 Blakely, D.W., Kozak, E., Sexton, G.A. & Somorjai, G.A. *J. Vac. Sci. Technol.* 1976, **13**, 1091.
66 Bridge, M.E., Comrie, C.M. & Lambert, R.M. *Surface Sci.* 1977, **67**, 393.
67 Prior, K.A., Schwaha, K. & Lambert, R.M. *Surface Sci,* 1978, **77**, 193.
68 Bozso, F., Ertl, G. & Weiss, M. *J. Catal.* 1977, **50**, 519.
69 Spencer, N.D., Schoonmaker, R.C. & Somorjai, G.A. *J. Catal.* 1982, **74**, 129.
70 Ertl, G., Weiss, M. & Lee, S.B. *Chem. Phys. Lett.* 1979, **60**, 391.
71 Bartholomew, C.H., Agrawal, P.K. & Katzer, J.R. *Adv. Catal.* 1982, **31**, 136.
72 Foord, J.S., Reed, A.P.C. & Lambert, R.M. *Vacuum* 1983, **33**, 707.
73 Foord, J.S. & Reynolds, A.E. *Surface Sci.* (in press).
74 Foord, J.S. & Lambert, R.M. *Suppl. Rev. Le Vide,* 1980, **201**, 211.
75 Johnson, S.W. & Madix, R.J. *Surface Sci.* 1981 **103**, 361.
76 Broden, G., Gafner, G. & Bonzel, H.P. *Surface Sci.* 1979, **84**, 295.
77 Kitson, M. & Lambert, R.M. *Surface Sci.* 1980, **100**, 368.
78 Goddard, P.J. & Lambert, R.M. *Surface Sci.* 1981, **107**, 519.
79 Foord, J.S. & Lambert, R.M. (unpubl. data).
80 Sinfelt, J.H. *Acc. Chem. Res.* 1977, **10**, 15.
81 Williams, F.L. & Nason, D. *Surface Sci.* 1974, **45**, 377.
82 McLean, D. *Grain Boundaries in Solids.* Clarendon, Oxford, 1957.
83 Buck, T.M. In *Chemistry and Physics of Solid Surfaces IV,* Vamselor, R. & Howe, R. (Eds), p. 437. Springer, Berlin 1982.
84 Seits, D.H. & Spicer, W.E. *Phys. Rev. Lett.* 1968, **20**, 25.
85 Ponec, V. *Adv. Catal.* 1983, **32**, 149.
86 Stoop, F., Toolemaar, F.J.C.M. & Ponec, V. *J. Chem. Soc. Chem. Commun.* 1981, 1024.
87 Vickerman, J.C., Christmann, K. & Ertl, G. *J. Catal.* 1981, **71**, 175 and references therein.
88 Somorjai, G.A. *Chemistry in Two Dimensions Surfaces.* Cornell UP, 1981.
89 Somorjai, G.A. *Adv. Catal.* 1977, **26**, 2.
90 Spencer, N.D. & Somorjai, G.A. *Rep. Prog. Phys.* 1983, **46**, 1.
91 Goodman, D.W., Kelley, R.D., Madey, T.E. & Yates, J.T. Jr. *J. Catal.* 1980, **63**, 226.

92   Kelley, R.W. & Goodman, D.W. In *The Chemical Physics of Solid Surfaces and Heterogeneous Catalysis*, Vol. 4, King, D.A. & Woodruff, D.P. (Eds). Elsevier, Amsterdam, 1982.
93   Grant, R.B. & Lambert, R.M. *J. Chem. Soc. Chem. Commun.* 1983, 662.
94   Cox, M.P., Ertl, G., Imbihl, R. & Rustig, J. *Surface Sci.* 1983, **134,** L517.

3   Catalysis by zeolites

M.S. Spencer

1   Introduction and history

Although the use of zeolites as catalysts in industrial chemical processes is only about 20 years old, zeolites have been recognized as a distinct family of minerals for more than two centuries. The word 'zeolite', which was coined in 1756, some 80 years before Berzelius invented the word 'catalysis', comes from the Greek and means 'a boiling stone'. A group of hydrous silicates, the zeolites, were found to swell and froth on heating under a blow pipe. This property derived from the

64

Table 3.1. Zeolite history (expanded from reference 7)

1756	Discovery and naming of first natural zeolite, stilbite
1825	Discovery of natural levynite
1842	Discovery of natural faujasite
1862	First zeolite synthesis (levynite)
1864	Discovery of natural mordenite
1870–88	First ion exchange studies with zeolites
1890	Discovery of natural erionite
1929	Potential as strong acids described (Pauling)
1930–34	First zeolite structure determinations
1932	Zeolites described as 'molecular sieves'
1942–45	Quantitative separations by molecular sieving
1948	First purely synthetic zeolite made
1948	Synthesis of mordenite
1949	Preparation of acid forms of zeolites
1956–64	Synthesis of zeolites A, X and Y (Union Carbide)
1962	Introduction of zeolite-based cracking catalysts (Mobil)
1971–72	Highly-siliceous zeolites (ZSM-5, ZSM-8) synthesized (Mobil)
1975	ZSM-5 catalysts used in ethyl benzene production
1978	ZSM-5 catalysts used in oil dewaxing
1978	Structures of ZSM-5 and ZSM-11
1980	High resolution electron microscopy and NMR applied to zeolites
1985	Methanol to gasoline plant (ZSM-5 catalyst) due to start up

special structural feature of zeolite minerals, for the 'boiling' was indeed due to water, expelled from the pore system which is part of the zeolite crystal structure. Some of the significant dates are given in Table 3.1.

Table 3.2. Some industrial processes in which zeolite catalysts are used

Process	Zeolite catalyst*
Oil Refining	
Cracking of middle and heavy oils (FCC)	REY(HY or REX)[†]
Hydrocracking of middle and heavy oils	Pd/HY
Paraffin isomerization	Pt or Pd/Hmordenite
Selective hydrocracking	NiS/Herionite
Heavy oil dewaxing	HZSM-5
Increase octane no. of catalytic reformate (M-forming)	HZSM-5
Petrochemicals	
Xylene hydroisomerization	Ni/HZSM-5
Xylene isomerization	HZSM-5
Ethylbenzene production	HZSM-5 (modified)
Toluene disproportionation	HZSM-5 (modified)
p-Methylstyrene production	HZSM-5 (modified)
Methanol conversion to gasoline	HZSM-5

*H, acidic (i.e. H[+]) form of zeolite.
 RE, rare earth exchanged zeolite.
[†]Zeolite minor component in a matrix.

The more important of the large-scale chemical and oil refining processes in which zeolite catalysts are used are listed in Table 3.2. Two features should be noted here: all the processes are in the oil refining and petrochemicals areas; of the 200 or so known zeolites of natural and synthetic origin, very few have found use as catalysts. The reasons for both these features will emerge later in this review. Catalytic cracking (fluid catalytic cracking, FCC) remains by far the most important of zeolite-catalysed processes and it is indeed in many ways the most important of all catalytic processes. The introduction of zeolite catalysts into FCC processes brought about a startling rejuvenation of a mature, established process. The other important recent discovery has been the synthesis of highly-siliceous zeolites. So far only one of these, Mobil's ZSM-5, has been used commercially as a catalyst, but it is already used in a wider variety of processes than any other zeolite catalyst.

Almost all conventional (i.e. non-zeolitic) catalysts, when looked at with a sufficiently high magnification, resemble a heap of builders' rubble: some common shapes recur (bricks or microcrystals) but the solid pieces are assembled randomly. Catalysis occurs on the surface of a heterogeneous catalyst so a large surface area is needed in an industrial catalyst to reach a useful rate of production. This is achieved by the use of very small, solid particles, which may be only a few Angstroms across. In contrast, the microview of a zeolite catalyst is very different. The zeolite crystals are much larger, typically one micron size, so their external surface area is relatively small. The appearance of a zeolite crystal resembles precision engineering on a microscale, for the zeolite crystals are themselves porous as an essential part of the crystal structure and catalysis takes place mainly within the zeolite pores. Some or all of the faces of the zeolite crystal contain regular rows of holes of uniform size leading to the internal pores. The diameters of the pores, defined closely by the zeolite crystal structure, are of molecular dimensions, hence the 'molecular sieving' action, of use in separations and of importance in catalysis.

This review is restricted mainly to the chemistry of those industrial processes which use zeolite catalysts. Some of the many other reactions which can be catalysed by zeolites are included to illustrate general features. Lists of books and reviews covering all aspects of zeolites[1–19] and especially catalysis by zeolites,[20–40] are given at the end of the review.

Zeolites also find industrial applications in many areas other than catalysis (e.g. ion exchange, gas separation, drying gases and liquids, detergent formulations) but these will not be described here. Information can be found in the general references.[1–19]

2   Properties of zeolites

2.1   Crystal structure of zeolites

Zeolites form one group in the range of silicate minerals, with the distinguishing characteristic of intracrystalline porosity. All silicate minerals, including pure silica, can be regarded as constructed of SiO_4 tetrahedra, which may be separate or share 1, 2 or 3 oxygen atoms in the structure.[4, 7, 17, 41] A wide variety of metal ions can be incorporated in these structures, either replacing Si in the framework or as separate cations (frequently hydrated) in gaps in the structure; indeed, for many structures these metal ions are essential. In zeolites all the SiO_4 tetrahedra are linked only via single O atoms (i.e. corner-sharing rather than side- or face-sharing) and aluminium is essentially the only metal atom which can readily replace Si in the framework. No doubt it is more difficult to prevent re-organization, or even lattice collapse, in an open structure than a closed structure, so the crystal structure requirements for zeolites are more severe. Zeolite structures with germanium in place of silicon or gallium in place of aluminium have been prepared[3] but they are of lower stability and little interest for catalysis. Many claims have been made for the incorporation of other metals in zeolite frameworks but there is little evidence that this can be done at more than trace levels. The catalytic activity of these materials can be attributed to Al impurities.[42] However, new materials with zeolite-like frameworks have been made in the Al_2O_3/P_2O_5 system.[43] Although zeolite frameworks are built only from SiO_4 tetrahedra, with some Al substitution, there are few constraints other than size on the materials which fill the pores in the framework.

The pore systems of different zeolites can consist either of separate one-dimensional tunnels or of intersecting tunnels in two- or three-dimensional arrays.

Replacement of the tetravalent silicon in the zeolite structure by the trivalent aluminium requires a negative charge on the Al atom. Thus, the framework has a net negative charge which is compensated by the cations in the zeolite pores, e.g.

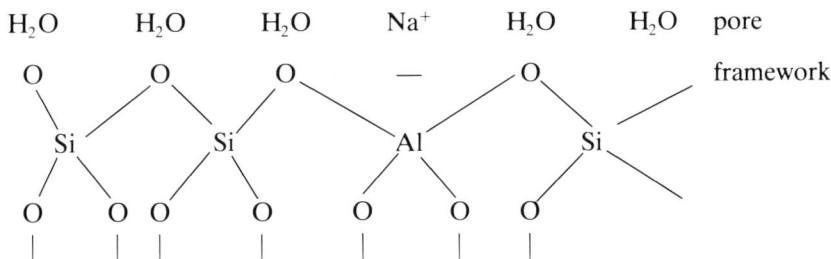

Bi- and tri-valent metal cations and cations formed from organic bases, e.g. $(CH_3)_4N^+$, can also function as counter ions. The size of zeolite unit cells, corre-ponding to the repeat distances in the structure, can be surprisingly large for an inorganic material. Thus, the faujasite structure, which the synthetic zeolites X and Y also have, has a cell volume of 15014 $Å^3$. These zeolites differ in composition: the Si/Al ratio is 1.0–1.5 for X, $> 1.5-\sim 3$ for Y and ~ 2.25 for faujasite, which occurs naturally with Na, K, Mg and Ca cations. Of importance for catalysis, the void volume is 47% of the total volume and part of this consists of cavities about 13 Å in diameter linked by ports of 7.4 Å diameter formed by rings of 12 SiO_4 tetrahedra.[4]

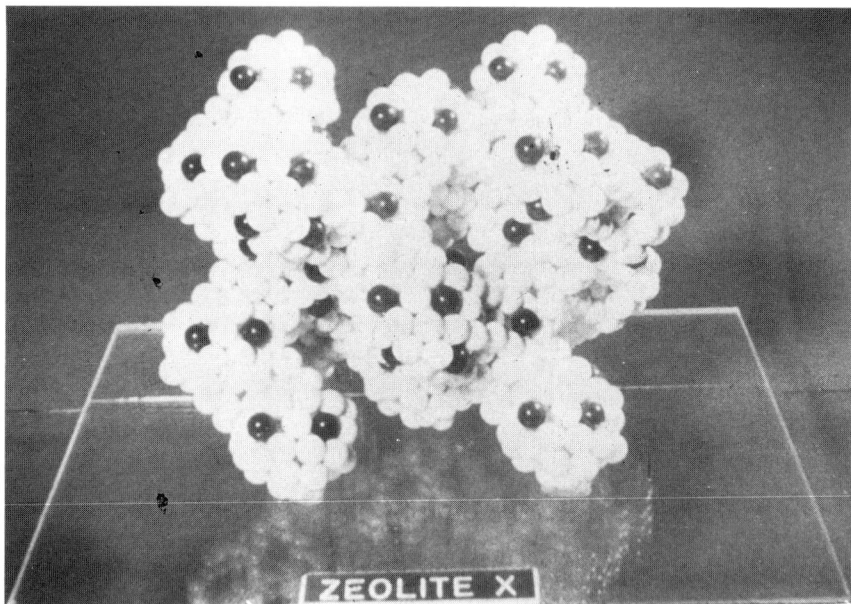

Fig. 3.1. Space-filling model of zeolite X. Note apertures (12-O ring windows) leading to internal cavities (supercages) in which catalysis can take place. (Photograph by courtesy of Union Carbide Corporation.)

A model of zeolite X (Fig. 3.1) shows these features. If distorted from a circular shape, the 12 SiO_4 ring will be of different dimensions, as in mordenite (Fig. 3.2), where the apertures are 6.7×7.0 Å. Both these structure types are classed as large-pore zeolites. The ports in ZSM-5 and ZSM-11 are built of 10 SiO_4 units (Fig. 3.3) and so are classed as medium-pore zeolites. The 10-ring cavities shown in Fig. 3.3 are linked to sinusoidal 10-ring tunnels in ZSM-5 and linear 10-ring tunnels in ZSM-11. Small-pore zeolites are characterized by 8-ring or 6-ring ports, e.g. the largest ports in zeolite A are 8-rings with apertures of 4.2 Å. Small-pore zeolites are widely used for separation and drying processes but most catalytic applications need medium- or large-pore zeolites. Many zeolites have

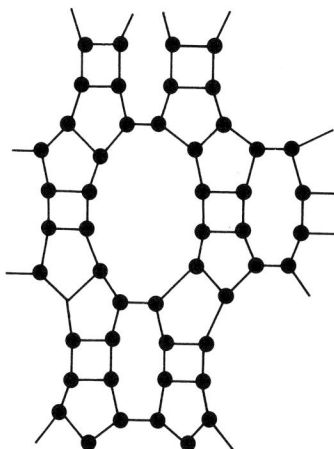

Fig. 3.2. Line diagram representing the structure of the zeolite mordenite. Silicon and aluminium atoms are at the line junctions, oxygen atoms at the mid-points of the lines. The main aperture is a distorted 12-O ring window.

small as well as large pores in the structure, e.g. mordenite has 8-ring windows, 2.9×5.7 Å in size, perpendicular to the large tunnels, but the larger tunnels are almost always the more significant in catalysis. The dimensions and shapes of the pore systems of zeolites can be deduced from the crystal structure or, especially in the case of zeolites with unknown crystal structures, the sorption of a range of molecules of graded sizes indicates size-limited access to the interior of the zeolite. The most vivid demonstration of zeolite pore structure is obtained by high-resolution electron microscopy[17,44] shown in Fig. 3.4.

The precision of a pore system defined by the atomic framework gives a very effective molecular sieving action,[4,7] of great importance in determining selectivity in catalysis (section 3). The molecular sieving action arises from the exclusion of molecules too large for the pore system (or, sometimes in catalysis, the inclusion of molecules too large to escape). Large differences in diffusion rates can be found even among different molecules which have access to a given zeolite pore system, with diffusion coefficients of similar molecules differing several orders of magnitude,[23] e.g. the diffusion coefficient of *n*-dodecane in erionite is about 10^{-11} cm^2 s^{-1} whereas that of a smaller molecule, *n*-octane, is only about 10^{-13} cm^2 s^{-1}. Activation energies for diffusion are also strongly dependent on both diffusing molecule and zeolite pore structure. Some examples[31] of the diffusion of aromatics in zeolite NaZSM-5 are given in Table 3.3. As in other fields (gases in dense zeolites and glasses or carbon dyes in fibres and polymers), the activation energy rises with closer approach between the void dimensions and those of the molecular configurations. However, some caution is needed in the

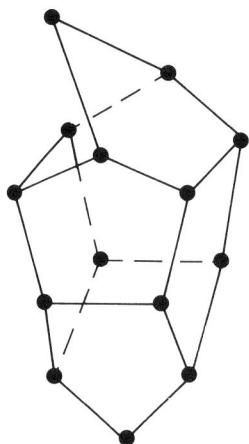

8–5 ring unit ZSM–5

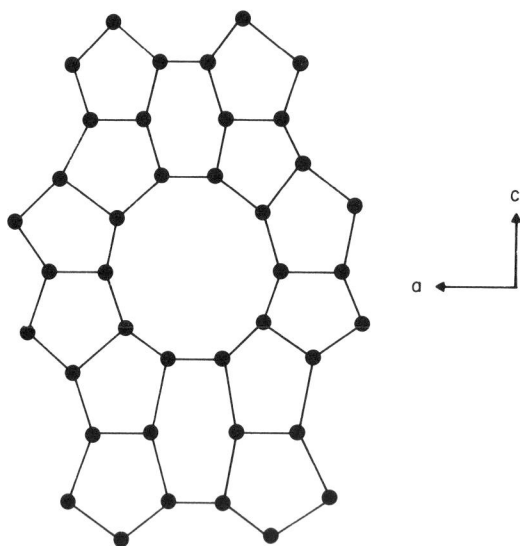

ZSM–5 in the ac plane

Fig. 3.3. Line diagram representing the structure and building unit of the zeolite ZSM-5. The main aperture is a 10-O ring window.

assessment of molecular sieving effects, for two factors militate against rigorous exclusions: thermal vibration, both in sorbate molecules and in the zeolite frame- work, allows the access at high temperatures of molecules which are just excluded at lower temperatures. It is also being realized increasingly that the Si–O bonds in zeolite structures can be labile. Facile exchange of oxygen between water or oxygen gas and the zeolite framework has been observed. Thus some widening, if

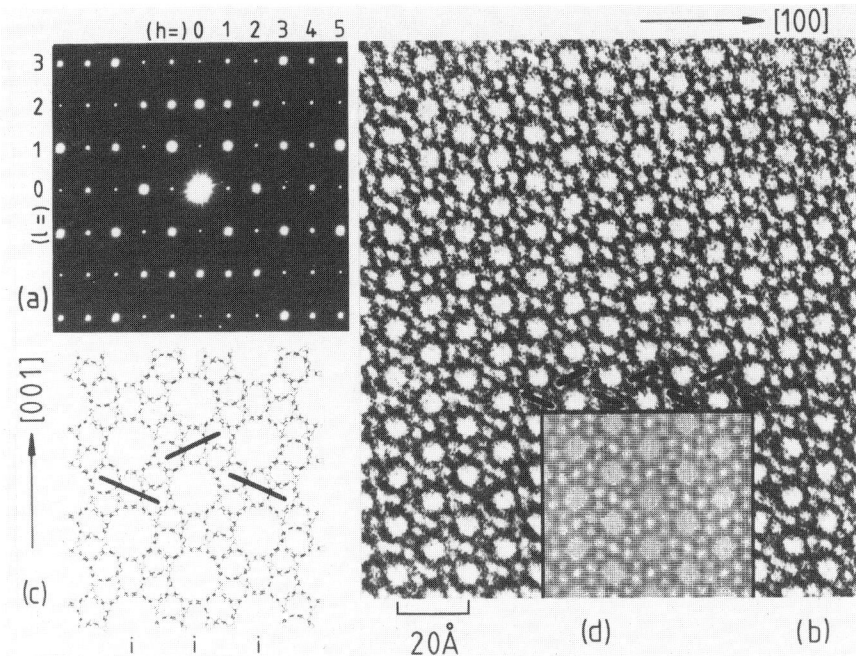

Fig. 3.4. The structure of zeolite ZSM-5 displayed by high-resolution electron microscopy. The diffraction pattern (a), real-space image (b), schematic drawing (c) (compare with Fig. 3) and compacted image for specimen thickness of 5.1 nm (d), all viewed along the *b* axis. The regular array of 10-O ring tunnels is clearly seen. (Photograph by courtesy of Professor J.M. Thomas.)

Table 3.3. Activation energies for diffusion of aromatics in NaZSM-5, from sorption studies between 250°C and 350°C[31]

Aromatic molecule	Activation energy (kcal mol^{-1})
o-xylene	9
m-xylene	14
t-butylbenzene	12
1,2,4-trimethylbenzene	14
1,3,5-trimethylbenzene	19

only temporary, of zeolite pores may take place during catalysis, especially at high process temperatures in the presence of steam, as in catalytic cracking.

2.2   Acidic properties of zeolites

The acidic form of the zeolite is required for most catalytic applications of zeolites. The possibility that aluminosilicate minerals could be strongly acidic was first suggested by Pauling in 1929 when he pointed out the formal similarity between the AlO_4^- group and the perchlorate ion ClO_4^-. Silica itself is essentially non-acidic (although impurities in commercial silicas give acidic properties), for the ionization of a silanol group is difficult:

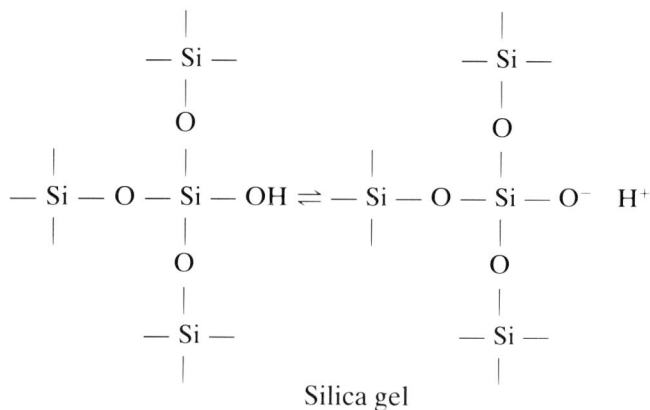

$$
\begin{array}{ccc}
\quad|\qquad\qquad\qquad\qquad\quad| \\
-\,Si\,-\qquad\qquad\qquad\ -\,Si\,- \\
\quad|\qquad\qquad\qquad\qquad\quad| \\
\quad O\qquad\qquad\qquad\qquad\ O \\
|\quad\quad|\qquad\qquad\quad|\quad\quad| \\
-\,Si\,-\,O\,-\,Si\,-\,OH \rightleftharpoons -\,Si\,-\,O\,-\,Si\,-\,O^-\ \ H^+ \\
|\quad\quad|\qquad\qquad\quad|\quad\quad| \\
\quad O\qquad\qquad\qquad\qquad\ O \\
\quad|\qquad\qquad\qquad\qquad\quad| \\
-\,Si\,-\qquad\qquad\qquad\ -\,Si\,- \\
\quad|\qquad\qquad\qquad\qquad\quad|
\end{array}
$$

Silica gel

In contrast, the removal of a proton from a silanol group adjacent to an aluminium atom in a zeolite gives the AlO_4^- group:

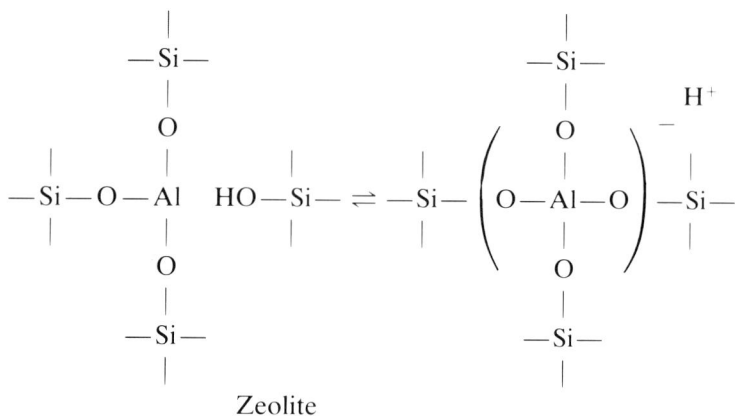

$$
\begin{array}{cc}
\quad|\qquad\qquad\qquad\qquad\qquad\quad| \\
-Si-\qquad\qquad\qquad\qquad\ -Si- \\
\quad|\qquad\qquad\qquad\qquad\qquad\quad|\qquad H^+ \\
\quad O\qquad\qquad\qquad\qquad\qquad\ O \\
|\quad\quad|\qquad\qquad|\quad\quad|\quad\ \ \ \ |\qquad\quad| \\
-Si-O-Al\quad HO-Si- \rightleftharpoons -Si-\left(O-Al-O\right)-Si- \\
|\quad\quad|\qquad\qquad|\qquad\ \ |\quad\ \ \ \ |\qquad\quad| \\
\quad O\qquad\qquad\qquad\qquad\qquad\ O \\
\quad|\qquad\qquad\qquad\qquad\qquad\quad| \\
-Si-\qquad\qquad\qquad\qquad\ -Si- \\
\quad|\qquad\qquad\qquad\qquad\qquad\quad|
\end{array}
$$

Zeolite

The residual negative charge is now distributed over the AlO_1 unit instead of one O and from simple electrostatics less work is required to remove the proton, i.e. it is a stronger acid.

So far acidity has been equated with the removal of a proton to some base, i.e. with the Brønsted form of acidity. Dehydration of the acidic forms of zeolites converts some at least of the Brønsted acid sites to Lewis acid sites. There is much controversy about the nature of Lewis acid sites: the simplest model of the Lewis acid site is one with trivalent Al in the surface, e.g.

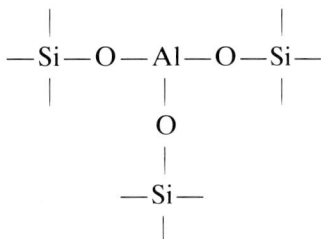

$$-\text{Si}-\text{O}-\text{Al}-\text{O}-\text{Si}-$$
$$\underset{\displaystyle \text{O}}{\mid}$$
$$\underset{\displaystyle -\text{Si}-}{\mid}$$

but there is evidence that some Al is removed from the framework during dehydration.

The total acidity of a zeolite catalyst depends on both the concentration of acidic sites and the strength of the individual sites. For high silica zeolites the acidic sites are well separated and ion-exchange capacity, IR absorption of OH and catalytic activity in hexane cracking correspond with aluminium content.[45] More recent work[42] with ZSM-5 has shown this holds for Si/Al ratios over nearly four orders of magnitude. Some results for ZSM-5 and HFU-1[46] are shown in Table 3.4. At higher aluminium concentrations, however, a mutual interference between acidic sites decreases the acidic strength of each site, so the high alumina zeolites have more acidic sites but of lower strength than the high silica zeolites. Maximum overall acidity is often found for Si/Al ratios in the range 5–20.

The influence of zeolite structure on zeolite acidity can best be understood if the zeolites are regarded as rigid ionizing solvents.[6,19,30] Acid strength is a measure of the energy required to remove a proton (assuming a Brønsted type of acidity) from the acid, but in practice the proton is added to a base to form a cation. As the

Table 3.4. Acid site concentrations in highly-siliceous zeolites (from reference 46)

Technique	Acid site concentration (no g^{-1} zeolite)	
	Zeolite HZSM-5	Zeolite HFU-1
Al content	3.2×10^{20}	8.4×10^{20}
CH$_3$NH$_2$ chemisorption	2.2×10^{20}	7.1×10^{20}
Pyridine chemisorption	2.3×10^{20}	$7.2 \times 10^{19*}$
Poisoning of but-l-ene	3.1×10^{20}	5×10^{20}
isomerization	(pyridine)	(NH$_3$)

*Low value because pyridine adsorbed on the exterior surface only of HFU-1, i.e. no access to internal acidic sites.

zeolite pores are of molecular dimensions the oxide ions of the framework (the lining of the pores) provide a rigid solvation sheath for any cation formed within the zeolite. Thus, the pore system of a zeolite can influence a catalysed reaction both by influencing transport to and from the active site and by controlling what happens at the active site.

3   Manufacture of zeolites

Zeolite manufacture is carried out on a large scale, e.g. the total worldwide consumption of zeolites in catalytic cracking is about 40000 ton per year.[11] A commercial zeolite plant is shown in Fig. 3.5. Zeolites are synthesized[4,13] by hydrothermal processes under alkaline conditions. Some zeolites, e.g. A, X and Y, can be made at 100°C or less and so conventional vessels can be used, but the newer, more siliceous zeolites require higher temperatures, either to make the zeolite in an acceptable time or indeed to make it at all. Large autoclaves are necessary for these processes. An autoclave used by Laporte Industries for the small-scale synthesis of high silica zeolites is shown in Fig. 3.6. Suitable sources of silica and alumina, together with appropriate cations, react, usually to give initially an amorphous gel which later recrystallizes into the appropriate zeolite. The material formed depends critically on many experimental parameters, such as

Fig. 3.5. Plant for the manufacture of zeolites. An aerial view of Union Carbide's plant at Mobile, Alabama, USA. (Photographed by courtesy of Union Carbide Corporation.)

the sources of silica, e.g. sodium silicate or colloidal silica, composition of the reaction mixture, seeding, ageing times at different temperatures, degree of agitation and time of reaction.[47] The organic bases used in the synthesis of high silica zeolites, e.g. tetra-isopropylammonium salts for ZSM-5, function partly as templates for the pore system but it is becoming clear this is not their only role. The synthesis usually results in a narrow range of crystal sizes, frequently about 1 μm in size. Crystal habits vary widely: Figs 3.7 and 3.8 show crystals of zeolites Y and ZSM-5. Zeolite FU-1 is different[18] from most zeolites in crystallizing in thin sheets (Fig. 3.9).

Fig. 3.6. An experimental autoclave for the synthesis of high-silica zeolites at temperatures above 100°C. (Photograph by courtesy of Laporte Industries Ltd.)

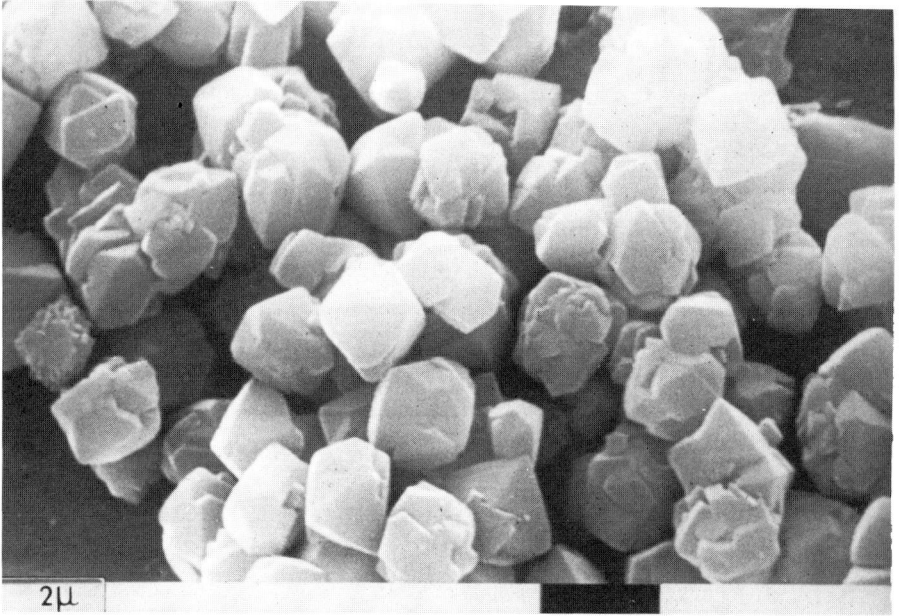

Fig. 3.7. Electron microscopy picture of crystals of zeolite Y. (Photograph by courtesy of Laporte Industries Ltd.)

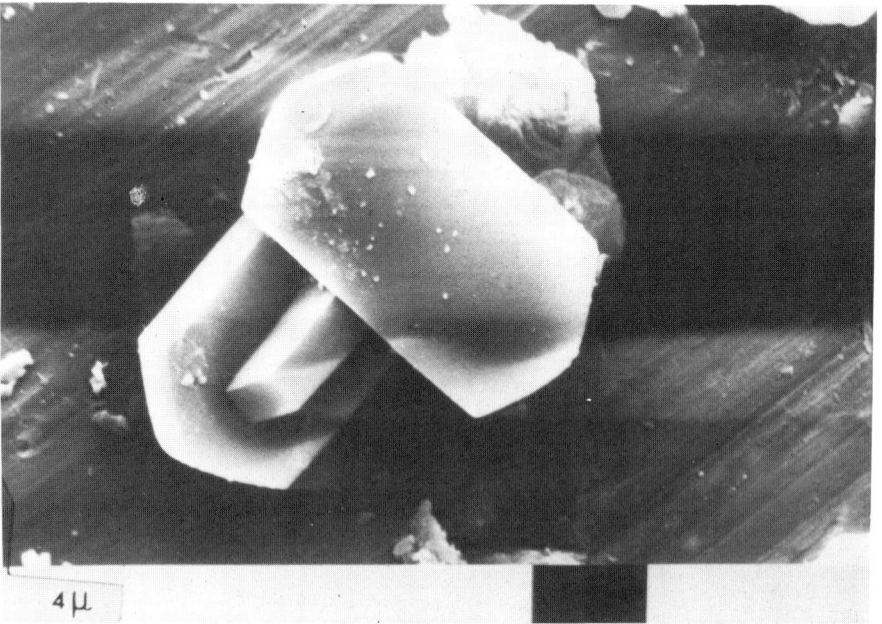

Fig. 3.8. Electron microscopy picture of crystals of zeolite ZSM-5. (Photograph by courtesy of Laporte Industries Ltd.)

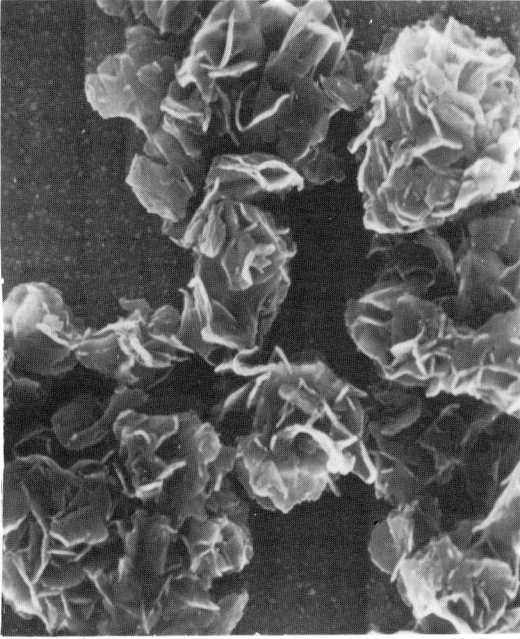

Fig. 3.9. Electron microscopy picture, at 10 000× magnification, of agglomerates of laminar crystals of zeolite FU-1 (after reference 18).

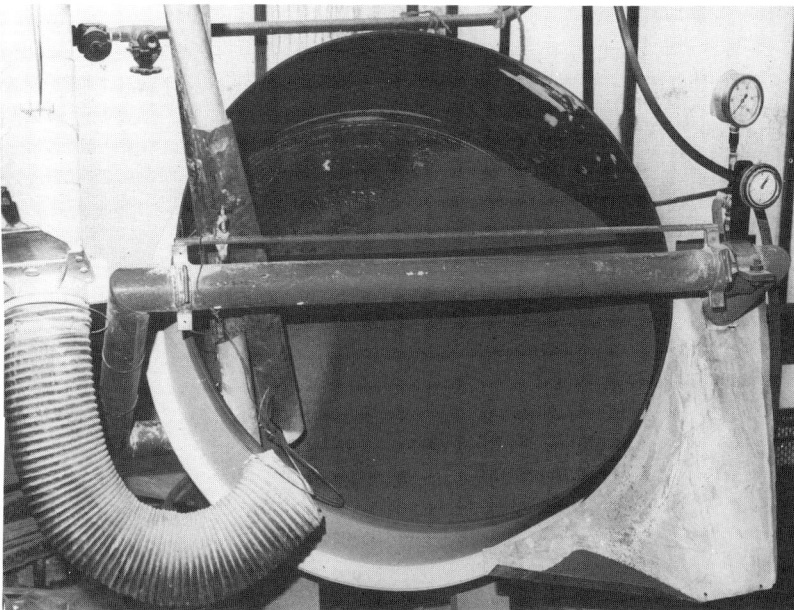

Fig. 3.10. Dish granulator used to make granules of zeolites for drying, separation or catalytic applications. (Photograph by courtesy of Laporte Industries Ltd.)

Once synthesized the zeolite crystals, when required for catalysts, have to be formed into suitable particles by the usual process of pelleting, granulation or extrusion (Fig. 3.10). A suitable binder, e.g. γ-alumina, is almost always necessary. Zeolites for catalytic cracking are incorporated in a matrix (section 5).

For most catalytic applications, the acidic forms of the zeolites are required, but the zeolites as synthesized contain cations, usually alkali metal or tetra-alkyl ammonium. Direct treatment with acid is usually unsuccessful as it causes structural collapse of high-alumina zeolites and it fails to remove large organic cations from high-silica zeolites. Several techniques are used, depending on the zeolite and the final forms required.[4,7,24]

1   Conversion by ion exchange from the sodium to the ammonium form, followed by calcination to remove ammonia.

$$Z^-Na^+ \rightarrow Z^-NH_4^+ \rightarrow Z^-H^+ + NH_3$$

2   Conversion by ion exchange to a bivalent or trivalent cation form, especially rare earth salt, followed by calcination to remove water. In principle the reactions are, e.g.

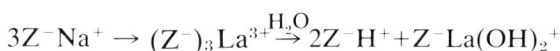

$$3Z^-Na^+ \rightarrow (Z^-)_3La^{3+} \xrightarrow{H_2O} 2Z^-H^+ + Z^-La(OH)_2^+$$

but the structural details are usually more complex.

3   Calcination in air is often effective in the oxidative removal of nitrogeneous bases, e.g.

$$Z^-(CH_3)_4N^+ + air \rightarrow Z^-H^+ + N_2, CO_2, H_2O$$

4   Ion exchange with transition metal ion, followed by reduction of the ion to the metal, is useful in forming a metal/H zeolite catalyst:

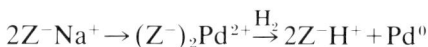

$$2Z^-Na^+ \rightarrow (Z^-)_2Pd^{2+} \xrightarrow{H_2} 2Z^-H^+ + Pd^0$$

Frequently, combinations of these techniques have to be used.

4   Fundamental aspects of catalysis by zeolites

The good industrial catalyst possesses three virtues: activity, selectivity and life. The special properties of zeolites are peculiarly appropriate to their catalytic applications in giving these properties.

4.1   Activity of zeolite catalysts

In most catalytic uses of zeolites their acidity is the fundamental basis of catalytic activity. Although a process like the catalytic cracking of distillate oils involves a complex network of reactions, the overall rate is determined largely by the rate of formation of hydrocarbon cations, which in turn is a function of the number and strength of the acidic sites. This is also true of other complex reactions such as the conversion of methanol to gasoline. The advantages of zeolites over other acidic materials, e.g. clays, can be seen in terms of the properties described in the previous section: each Al atom in the structure provides a strongly acidic site, in contrast to the amorphous silica–alumina where a small fraction only of the Al atoms are sufficiently active. Furthermore, the 'solid solvent' effect of the zeolite pore structure stabilizes the cationic reaction intermediates. Poutsma[48] has emphasized the need to consider the solvation of the carbenium ions formed in the catalysis of hydrocarbon reactions. 'This solvation must be geometrically dependent on the rigid framework structure rather than effectively continuous as in solution . . . predictions of maximum activity in terms of zeolite structure might need to consider not only O–H acidity and density but also its distribution with respect to availability of adjacent carbenium ion adsorption sites'. At present this view can be taken no further than general, qualitative considerations. As long as access is not critically impeded, solvation of cationic intermediates should improve with 'fit' in the pore structure, i.e. better solvation should be found in small pores than in large pores and solvation should be negligible on amorphous silica–aluminas and the exterior surface of zeolites.

The differences in catalytic activity also depend on the ease or difficulty in forming carbenium ions from the hydrocarbon reactant. Hexane is difficult to ionize, so its cracking is several orders of magnitude faster over hydrogen zeolites than over amorphous silica–alumina, but the differences in rates of reaction of aromatic hydrocarbons, e.g. xylene, are much smaller because a carbenium ion can be formed readily by proton addition on both catalysts.

Although the interior surface of zeolite catalysts is the main source of catalytic activity there is evidence[18,49] that reactions also occur on the external surface.

4.2   Selectivity of zeolite catalysts

It is in the area of catalyst selectivity that the full subtlety of zeolite properties can be exploited. Csicsery[40,50] has distinguished three categories of shape-selective catalysis:

(a)   *Reactant selectivity*: occurs when only some of the reactant molecules can pass through the catalyst pores to the active site. The remaining molecules are too large to diffuse through the pores and so they do not react.

(b)   *Product selectivity*: occurs when, among all the product species formed within the pores, only those with small enough dimensions can diffuse out and appear as observed products. Bulky products, if formed, are either converted to less bulky molecules (e.g. by equilibration) or eventually deactivate the catalyst by blocking the pores.

(c)   *Restricted transition-state selectivity*: occurs when certain reactions are prevented because the corresponding transition state would require more space than is available at the active sites in the cavities. Neither reactant nor potential product molecules are prevented from diffusing through the pores and reactions requiring smaller transition states proceed unhindered.

The limited transport required for 1 or 2 to operate does not need to be a total exclusion. The wide variations in diffusion coefficients noted above can bring about reactant or product selectivity even when all molecules can pass through the zeolite pore system.

Examples of all three types of selectivity have been found in reactions of industrial importance, but it is often difficult to establish the origin of selectivity under any given set of conditions. Reactant selectivity was used in the selective hydrocracking of n-paraffins by the small-pore zeolite, H-erionite.[23] Iso-paraffins and cyclic hydrocarbons were excluded from the acidic sites. In the conversion of methanol to gasoline (section 6.5), product selectivity limits the aromatics to C_6–C_{10}.

In M-forming, a process for upgrading naphthas for gasoline blending, the graduated selectivities of ZSM-5 for cracking various paraffin structures are used.[31] Over amorphous silica–alumina or other non-zeolite acidic catalysts normal paraffins are cracked less readily than branched paraffins, but this select-ivity is reversed with zeolite catalysts. The cracking ability of ZSM-5 for the

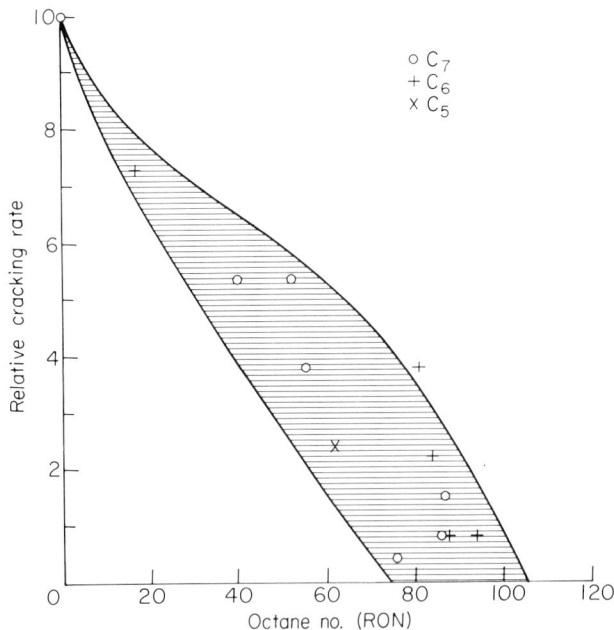

Fig. 3.11. Basis of M-forming. Selective cracking rate against research octane number (RON) for C_5–C_7 straight and branched chain paraffins. Reaction of a mixture gives an increase in octane number of product by the selective removal of low octane components (after reference 31).

various paraffin structures correlates inversely with their blending octane numbers (Fig. 3.11), thus making the catalyst what Weisz[31] has called an 'intelligent' octane number improver. In the same process, alkylation of aromatics with the cracked fragments also occurs, so increasing the liquid yield. Contrary to the conventional relative alkylation activities of benzene and toluene, benzene reacts preferentially in ZSM-5. This results in a desirable lowering of benzene/ toluene ratios in the product. The reversal of selectivity for paraffin cracking observed in zeolites was first attributed to reactant selectivity, i.e. it was suggested that although normal paraffins were less reactive than iso-paraffins, their faster rates of diffusion through medium-pore zeolites gave faster rate of cracking. More recent work[31,51] has shown that relative activation energies do not fit this explanation and, furthermore, the cracking rate constants are essentially unchanged over two orders of magnitude of zeolite crystal size. Selectivities are therefore not controlled by diffusive transport but by the intrinsic reaction kinetics, i.e. transition-state selectivity.

Another example of transition-state selectivity is found in the xylene isomerization process.[31,52,53] The more constrained the space is around a catalytic site, the more favoured monomolecular events will be over bimolecular events. This is believed to be the main cause of the high selectivity of the ZSM-5 catalyst in xylene isomerization. The desired process, isomerization, is mono-molecular:

$$m-C_6H_4(CH_3)_2 \rightarrow p-C_6H_4(CH_3)_2$$

whereas the unwanted side reaction, disproportionation, is bimolecular:

$$2C_6H_4(CH_3)_2 \rightarrow C_6H_5CH_3 + C_6H_3(CH_3)_3$$

Experimental data on selectivity in xylene isomerization over four zeolite catalysts of different pore dimensions are shown in Fig. 3.12.

4.3   *Life of zeolite catalysts*

Catalyst life can be defined as the length of time for which the essential activity and selectivity of a catalyst are maintained. Many varied causes of decay can bring about failure of the catalyst, but for zeolite catalysts most can be put into three categories.

(a)   *Coking.* Most if not all catalysed organic reactions give eventually an involatile carbonaceous deposit on the catalyst. Catalyst activity is lost, and sometimes selectivity modified, by coke deposition at the active site, through the pore system and on the external surface of the zeolite crystal. Thus, both the catalysed reaction itself and mass transport within the zeolite can be affected. Rates of coking on different zeolites in the same reaction can vary by several orders of magnitude, as in methanol conversion to gasoline (section 6.5). The consequences of coking are also dependent on zeolite structure: blockage of one-dimensional

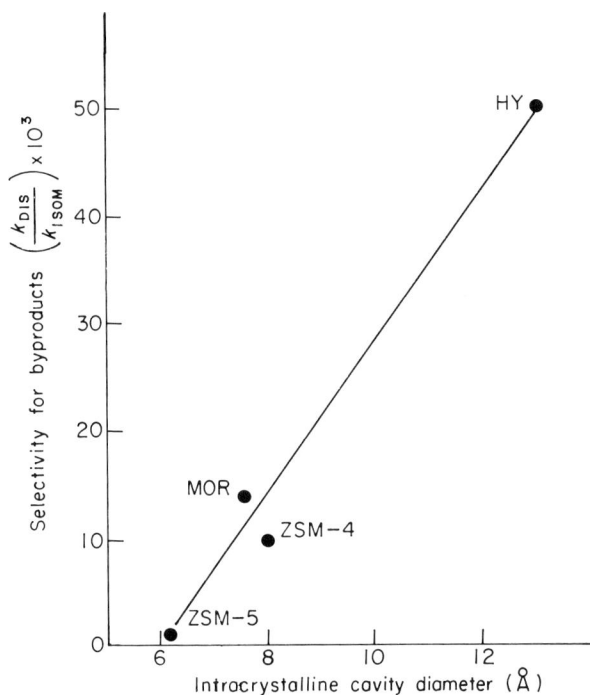

Fig. 3.12. Xylene isomerization. Selectivity (ratio of disproportionation to isomerization) as a function of the effective diameter of the zeolite pore (after reference 31).

tunnels prevents catalysis much more quickly than obstructions in a three-dimensional pore system. The high activity of acidic zeolites in both carbon–carbon bond formation and in hydrogen transfer leads generally to ready coke formation, but the space available in the zeolite pore structure for coke-forming reactions has an important influence on coking rates.[54]

Three ways of overcoming (or at best accommodating) coking are in general use. If possible, zeolite and process conditions are chosen so that the rate of coking is slow. Intermittent regenerations, by air oxidation, can then be acceptable if not too frequent. The methanol-to-gasoline process falls into this category. Intrinsically fast coking can be inhibited by the hydrogenation of unsaturated coke precursors. In hydrocracking a hydrogenation catalyst is incorporated in the zeolite (e.g. Pd in HY zeolite) and the cracking process is operated in a high pressure (100+bar) hydrogen, but this is expensive in plant, catalyst and hydrogen. In catalytic cracking (FCC), fast coking is accepted and the process engineered to allow fast regeneration by coke combustion.

(b) *Structure collapse.* In many non-zeolite catalysts, activity is lost in the sintering and loss of surface areas of microcrystallites of the catalytically-active phase. Metals supported in zeolite crystals can also be liable to sinter, so decreasing their effectiveness in preventing coking, but the zeolite crystals are far too large and stable for any sintering to be of practical consequence in zeolite catalysis. The analogous cause of decay in zeolite catalysts is collapse of the crystal structure of the zeolite. This occurs at high temperatures, and especially in the steam atmosphere produced by coke combustion. In the early stages the development of faults can cause pore blockage or modification to the active site, but at later stages complete collapse of the pore system takes place. Structure collapse is very dependent on zeolite type. Stability increases with Si/Al ratio and one advantage of the new highly-siliceous zeolites such as ZSM-5 is their great stability. Multi-valent ions, especially rare earth, also enhance structure stability.

(c)   *Poisons.* All catalysts are liable to fail from poisoning, i.e. loss of activity or selectivity as a consequence of traces of impurities either in the feed or in the catalyst itself. The remedies depend on the poisons but they are (i) to remove the poison or (ii) (if removal cannot be accomplished) to find some way of mitigating its effects. Thus, nitrogeneous organic bases are strongly sorbed at acidic sites in zeolites and so cause deactivation in cracking and other related reactions. Hydrodenitrogenation of the feedstock provides a way round the problem. Poisons can also promote zeolite catalyst decay from coking or structure collapse. Part of the difficulty of using residual feeds in catalytic cracking follows from the deposition of nickel and vanadium (present in residuum as asphaltenes) on the cracking catalyst. Possible remedies are de-asphalting the residual feed, treating the catalyst to inhibit dehydrogenation activity and more rapid catalyst replacement (section 4). The presence of salt in residual feeds can also be harmful (as can Na^+ ions from any source) in causing zeolite structure collapse. The amorphous matrix used in FCC catalysts to hold the zeolite acts as a 'sink' for sodium ions.

5   Catalytic cracking

5.1   The catalytic cracking process

Cracking processes found early use in oil refineries, for there was a frequent need for more light oil fractions (especially gasoline and diesel fuel) and less heavy oil fractions than could be obtained by distillation of crude oil. Early thermal, non-catalytic cracking processes have been largely displaced (except for coking processes) by catalytic processes[55]. The use of activated (by acid treatment) clay catalysts was discovered in the 1930s to give higher yields of gasoline of higher octane number.

Fluid catalytic cracking (FCC), which was developed[56] from earlier fixed-bed processes, is now operated on a large scale in refineries across the world (Table 3.5).

Catalytic cracking plants in the UK are listed in Table 3.6. Catalytic cracking is by far the most important catalytic process in oil refining, in terms of capacity,

Table 3.5. Catalytic cracking capacity in various countries, at 01.01.84. All capacities as barrels per calendar day[57]

Country*	Number of refineries	Crude oil distillation	Catalytic cracking[†]
Brazil	13	1301 400	315 400
Canada	28	1806 600	437 500
France	19	2670 400	250 300
Germany	27	2386 250	184 500
Italy	24	3050 100	271 200
Japan	45	5020 350	442 950
Mexico	9	1269 000	297 000
Netherlands	7	1551 500	115 000
Spain	10	1493 000	63 500
UK	16	2091 500	341 000
USA	225	15 930 000	4473 000
Total non-communist world	659	58 007 045	8467 380

*Major countries only included.
[†]Catalytic cracking capacity, as fraction of distillation capacity, varies between countries.

importance and catalyst use. A typical FCC unit can process some 7000 tons oil day^{-1} and to do this it uses a ton or more of cracking catalyst per day. Total world consumption of cracking catalysts is of the order of 150 000 ton year^{-1} with a value estimated[58] at $500 million year.$^{-1}$ A modern FCC plant is shown in Fig. 3.13.

An FCC plant (Fig. 3.14) consists essentially of three sections, in addition to the distillation required for product separation.[55] These are: the reactor, in which

Table 3.6. Catalytic cracking plants in the UK.[57] All capacities as barrels per calendar day

Company* and location	Crude oil distillation	Catalytic cracking[†]
England		
Esso, Fawley	304 000	76 500
Lindsey, Killingholme	190 000	35 000
Mobil, Coryton	145 000	36 000
Shell, Stanlow	256 000	52 000
Scotland		
BP, Grangemouth	173 000	17 000
Wales		
Amoco, Milford Haven	102 000	31 500
BP, Llandarcy	112 000	18 000
Texaco, Milford Haven	180 000 ⎫	75 000
Gulf, Milford Haven	103 000 ⎭	

*Only refineries with catalytic cracking plants are shown.
[†]All catalytic cracking plants are of the fluid catalytic cracking (FCC) type.

Fig. 3.13. 29 000 barrel per day fluid catalytic cracker in Atlantic Richfield's Philadelphia refinery. The plant incorporates the 'Ultra-Orthoflow' technology of Kellogg and Amoco. The disengager is sited above the regenerator and the riser reactor, at the side, connects the bottom of the regenerator to the disengager. (Photograph by courtesy of M.W. Kellogg Co.)

the oil feed is cracked; the stripper, where the products are separated from the catalyst; and the regenerator, needed to burn the coke from the catalyst. The equilibria of the desired cracking and aromatization reactions are favourable only above about 500°C. Under these conditions all acidic catalysts coke rapidly. The reactions are endothermic, so heat has to be supplied. These constraints are combined to advantage by the use of a fluidized, recycling catalyst bed in a continuous process, in which rapid and frequent catalyst regeneration by coke burning evolves heat to sustain the oil cracking. Early FCC plants had a separate vessel for the cracker reactor but the activity of modern zeolitic catalysts is so high that sufficient reaction occurs in the pipe (the 'riser reactor') which carries catalyst and oil from the bottom of the regenerator to the stripper. The oil meets the hot catalyst, essentially free of coke, as it falls through a valve at the bottom of the regenerator and passes into the riser reactor. Steam is used to aid the removal of

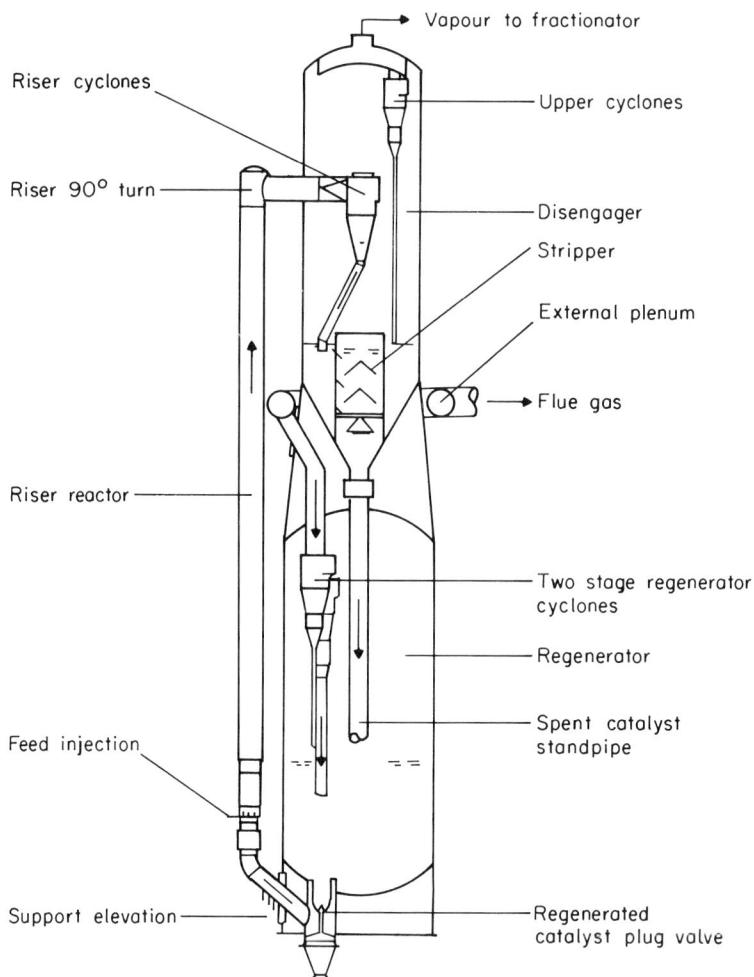

Fig. 3.14. Simplified flow sheet of catalytic cracking (FCC) plant.

products from the coked catalyst and a cyclone prevents fine catalyst particles from being entrained in the hydrocarbon products.

5.2   The chemistry of fluid catalytic cracking

Fluid catalytic cracking (FCC) was developed[56] for use with amorphous silica–alumina catalysts as microspheroids of about 50 μm diameter. The chemistry of the process,[27] which was not changed by the introduction of zeolite catalysts, depends on the formation and subsequent reactions of carbenium ions from the hydrocarbon feed, normally (but see section 5.4) oils from the distillation of crude oil either at atmospheric pressure (straight-run gas oil) or under reduced pressure (vacuum gas oil). High boiling products (cycle oils) are also included in the feed: these have a similar boiling range to fresh feed but have a much higher concentration of polynuclear aromatic compounds. Feeds and products consist of many thousands of individual compounds, so the process has to be described in terms of three groups of compounds present: paraffins, naphthenes (cyclo-paraffins) and aromatics.

Carbenium ions are formed from paraffins initially by the addition of an olefin (from thermal or catalytic cracking) to the Brønsted acid site on the catalyst:

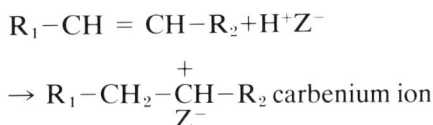

$$R_1-CH = CH-R_2+H^+Z^-$$

$$\rightarrow R_1-CH_2-\overset{+}{\underset{Z^-}{CH}}-R_2 \text{ carbenium ion}$$

Subsequent reactions of the carbenium ion—exchange reaction with another paraffin, scission to give a smaller carbenium ion+olefin, isomerization to give the more branched tertiary carbenium ion, hydride transfer—explain well the complex changes occurring during the cracking process. Recently the additional participation of 5-coordinated carbonium ions has been proposed.[59]

Naphthenes react in a similar way, but isomerization allows changes in ring size, e.g. 5-ring $\rightleftharpoons$ 6-ring, followed hydrogen transfer from the 6-ring to give aromatics. Alkyl aromatics also form carbenium ions, but on the ring rather than the side chain, so side chains, or naphthenic rings, split off as olefins.

5.3   Catalysts for fluid catalytic cracking

Some flexibility in operation is possible but most FCC units are operated to produce maximum gasoline yield with maximum octane number. The catalyst clearly must combine sufficient activity to give substantial conversion of the feed and suitable selectivity to high quality gasoline. There are other stringent constraints on the catalyst.[28,55] Coking must not be excessive, otherwise feedstock is wasted and only partial regeneration will be achieved. It has to withstand the high temperature (up to 800°C in some cases) and steam of the regenerator to allow regeneration without loss of activity. It must be stable against continuous thermal shock, for it passes continuously between high regeneration temperature and lower process temperatures (500–550°C). Poisoning by nitrogen, sulphur, vanadium and nickel compounds must be minimal. The physical properties must also include resistance to attrition, otherwise catalyst break-up will occur. Amorphous silica–alumina catalysts fulfilled all these requirements.

The introduction of X zeolites into FCC catalysts in 1962 depended on two achievements.[60] The first was the satisfactory preparation of acid forms of zeolite X, which was done via the ammonium salt, but the second, and more critical, was the achievement of sufficient hydrothermal stability so that the zeolite survived regeneration in the plant. Plank & Rosinski[60] found rare earth exchange to be the key here. The first zeolite catalysts contained 5–10% of rare-earth-exchanged zeolite X (REX) in an amorphous silica–alumina matrix. Higher concentrations of zeolite could not then be used because the FCC plant could not cope with the very high intrinsic cracking activity of acid zeolites. Of greater importance for the process was the small increase ($\sim 5\%$) in gasoline yield and the higher octane number of the gasoline by an increase in aromatics content. Although small, these advantages were of great value to refineries and in 2–3 years there was virtually a complete changeover to REX-containing catalysts in the USA. The gain in financial terms for the USA in 1967 was estimated[60] at \$250000000 per year.

5.4   Recent developments in fluid catalytic cracking

Subsequent developments have involved both catalyst and plant improvement. FCC plants were re-designed to take full advantage of the very high activity of zeolitic catalysts while avoiding over-cracking and loss of product. This led to the evolution of plants with the 'riser reactor' (Fig. 3.14), which in turn allowed the use of more zeolite, and of the more active zeolite Y in place of X, in the catalyst. Some data showing this advance is given in Tables 3.7 and 3.8. These show the complex interaction between catalyst and engineering design to give improved performance. The zeolite content of FCC catalysts can now be as high as 40%.[11]

Other advances in the catalyst include the addition of traces of platinum group metals to promote complete coke oxidation, i.e. to CO_2 rather than CO/CO_2 mixture. This in turn brings the advantages seen in Tables 3.7 and 3.8. The

Table 3.7. Performance of a representative FCC plant at constant air rate with different catalysts[61]

Performance	Catalyst				
	13% Alumina*	25% Alumina*	Low activity zeolite	High activity zeolite	High activity zeolite†
Conversion (%) to products of b.p. $< 221°C$	68	70	72	79	79
Yields:					
C_2 (wt %)	2.8	3.0	2.5	2.5	2.5
C_3 (vol %)	8.3	8.7	8.0	9.2	9.2
C_4 (vol %)	15.3	15.9	14.6	15.8	15.8
$C_5/221°C$					
gasoline (vol %)	52.0	52.7	56.7	62.8	64.2
heating oil (vol %)	27.0	25.0	23.0	16.0	16.0
Bottoms (vol %)	5.0	5.0	5.0	5.0	5.0
Coke (vol %)	5.9	5.9	5.9	5.9	4.8
Preheat temp. (°C)	366	379	388	321	266

*Amorphous silica–alumina catalysts.
†Zeolitic catalyst incorporating promoter for CO combustion.

Table 3.8. Improvement in performance of small FCC plant over 30 years[61]

Plant property	1951 Design	1967 Amorphous catalyst	Zeolite catalyst	High regen. temp.	Elim. recycle	Potential expansion
Fresh feed rate (bbl d^{-1})	3070	4070	5700	6000	9200	12500
Gasoline yield						
bbl d^{-1}	1700	2200	3900	4100	5400	—
% feed	55.6	53.2	68.4	68.3	58.7	—
Preheat temp. (°C)	—	338	388	366	427	371
Regen. Temp. (°C)	607	621	677	704	704	—
CO in flue gas (vol %)	10.0	10.0	11.0	3.0	1.0	—
Carbon on regen. catalyst (wt %)	0.7	0.8	0.3	0.05	0.05	—

preheat temperature can be reduced to give better energy conservation and gasoline yields are improved because of lower carbon levels on the regenerated catalyst.

There are two, sometimes conflicting, pressures for further changes in FCC:[39] an increase in the range of feedstocks used, especially to residual oils, and decreases in all aspects of pollution. Improved CO oxidation helps in both respects, for better regeneration enables the refiner to cope with heavier feeds which deposit more coke on the catalyst. A further problem with residual oils arises from the organic vanadium and nickel compounds they contain. Metallic nickel and vanadium oxides accumulate on the catalyst as it recycles through the plant, promoting dehydrogenation reactions and so causing excessive gas make and coking. Vanadium also affects catalyst cracking activity, due to the formation of low melting eutectics during the regeneration. Newer high zeolite catalysts, with low surface area matrices, suffer less from this trouble. The addition of antimony compounds to the feed is also used to passivate metals deposited on the catalyst. Even with these measures increased catalyst addition rates are necessary.

The control of SO$_x$ emissions for FCC plants has also received much attention recently. Catalyst and additives with sulphur transfer properties (e.g. the use of an alumina rather than a silica–alumina matrix) have become available and SO$_x$ decreases of 50% have been reported.[62]

Other new catalysts have been introduced recently which give significantly higher gasoline octanes from distillate and residual oils via increased aromatic and olefins content. These improvements are due to changes in zeolite and matrix composition.[63–66]

The catalytic cracking process is now barely recognizable as the descendent of

Table 3.9. Industrial processes in commercial use based on high-silica zeolite catalysts

Process	Objective	Main chemical/process characteristic
Distillate dewaxing	Light fuel from heavy fuel oil	Cracking of high molecular weight n- and mono-methyl-paraffins.
Lube oil dewaxing	Lube oils with low temperature flow point	
Xylene isomerization	High yield p-xylene production	High throughput; long cycle life; supression of side reactions.
Ethylbenzene production	High yield ethylbenzene production	
Toluene disproportionation	Benzene and p-xylene production	
Methanol-to-gasoline*	Methanol conversion to high grade gasoline	Synthesis of hydrocarbons only, in gasoline range (C_1 to C_{10}) including aromatics.

*Start up of first commercial methanol-to-gasoline plant due 1985.

Houdry's original process, due as much to changes in catalyst as changes in engineering. However, it is still doing essentially the same job of gasoline production, but much more efficiently and on a much larger scale.

6   High-silica zeolites

6.1   The new zeolites

Zeolites X and Y are used in other oil refining processes—for example, hydro-cracking[67]—but the most significant advances in the past decade have come from the synthesis and use of high-silica zeolites.[15–19,31,35–38] Unlike zeolites X and Y, which have essentially the same framework as the mineral faujasite, these new zeolites have no natural analogues. Two groups have been prolific in the synthesis of new high-silica zeolites, those at Mobil (ZK and ZSM series) and those at ICI/Edinburgh University (Nu, FU and EU series). As mentioned above (section 3), the preparation of these zeolites largely depends on the use of organic bases and high temperatures, $> 100°C$. Highly siliceous modifications of the more conventional zeolites can be made by various de-alumination techniques but, partly because of the decreased crystallinity of these materials, none has yet reached industrial use.

Although the various high-silica zeolites have different crystal frameworks, and so different pore structures, there are several characteristics common to most of these materials.

(a) Small to medium port sizes. Most of the high-silica zeolites have 8-ring or 10-ring ports, in contrast to the 12-ring ports of zeolites X and Y, and many have 5-ring units in the structure.

(b) The acidic sites, which have been correlated quantitatively with the Al atoms in the zeolite in ZSM-5[42,45] and Fu-1,[46] are very strongly acidic.

(c) Most of the structure of the high silica zeolites is essentially silica in the form of the zeolite lattice. This has greater stability than the aluminosilicate frameworks of, e.g. zeolites X and Y.

The Mobil zeolite, ZSM-5, is the only high-silica zeolite so far to achieve widespread use as a catalyst. The special qualities of ZSM-5 are discussed below in the description of the methanol to gasoline process.

A list of industrial processes which use high-silica zeolite catalysts, an extended version from Weisz,[31] is given in Tables 3.9 and 3.10. The processes in Table 3.10 (as yet not in full commercial use) have all been developed beyond the laboratory scale but are awaiting either further development or favourable economic circumstances. The basis of M-forming is discussed above in section 4.2. The other processes fall into four groups: dewaxing processes, aromatics, olefin and methanol conversion processes.

Table 3.10. Industrial processes based on high-silica zeolite catalysts but not yet in full commercial use

Process	Objective	Main chemical/process characteristic
M-forming	High yield; octane number increase in gasoline	Cracking dependent on degree of branching; aromatic alkylation by crack fragments.
Olefins-to-gasoline and distillate	High yield of gasoline or diesel fuel	Controlled oligomerization of olefins.
Methanol to olefins	Methanol conversion to C_2–C_4 olefins with minimum other hydrocarbons	For synthesis of hydrocarbons only; further reaction of olefins prevented.
p-Ethyltoluene	High yield of p-ethyltoluene from toluene and ethene	Minimum formation of o-, m-ethyltoluene and of higher products.
Toluene methylation	High yield of p-xylene from methanol and toluene	Minimum formation of o-, m-xylenes and higher methylbenzenes.
Cyclar	Conversion of LPG to high yield of aromatics	Bifunctional catalysis; high yield of aromatics from olefin intermediates.

6.2   Dewaxing processes

The two dewaxing processes operate in similar ways, but with different feedstocks. The Mobil middle distillate dewaxing process (known as MDDW) is aimed at increased diesel oil yield by the removal of wax molecules which would otherwise affect the cut-point of the finished product.[68] The wax molecules consist of n-paraffins and monomethyl-paraffins and these hydrocarbons are cracked faster in ZSM-5 catalysts than the more branched isomers (section 4.2). The ZSM-5 catalyst takes the wax content of the feedstock and cracks it to high-octane gasoline and a small amount of gas. The extent of improvement is controlled by adjustment of reactor temperature and/or liquid space velocity. The alternative to catalytic dewaxing is solvent dewaxing but this has always been too expensive for commercial use.

The MDDW process had its first commercial test in 1974 and recently substantial improvements were announced.[69] With the new catalyst, reactivation of the catalyst is much less frequent and product yield and quality have improved. The process can now cope with heavier feeds, e.g. catalytically cracked gas oils.

Solvent dewaxing is used extensively for lubricating oil manufacture, but catalytic dewaxing, again using a ZSM-5 catalyst, has the economic advantage of much lower capital cost.[68,69] Very low pour points can be achieved which are impossible on conventional solvent dewaxers because of thermodynamic limits. The full range of lubrication oil grades can be dewaxed by the Mobil process (MLDW), which differs from the distillate dewaxing process in the use of two reactors. Dewaxing occurs over the ZSM-5 catalyst in the first reactor and a conventional hydrofining catalyst is used in the second reactor to saturate olefins and stabilize the dewaxed lube product.

6.3   Aromatics conversion processes

Aromatics processing covers the xylene isomerization, toluene disproportionation and aromatics alkylation with ethylene (ethylbenzene and p-ethyltoluene production) processes. The main characteristic brought about by the use of high-silica zeolites in these processes is that of high selectivity. In xylene isomerization, a mixed xylene feed is brought to a near-equilibrium mixture of the isomers so that the commercially-desirable components (p-xylene and, to a less extent, o-xylene) can be removed and the residue recycled. Selectivity is essential in two ways: other reactions of xylenes, e.g. disproportionation to benzene and trimethyl benzene, must be minimized and ethylbenzene, always present in mixed xylenes, must be converted either to xylenes or benzene. Variations on the xylenes isomerization process (largely different reaction conditions) using catalysts based on ZSM-5 fulfil these requirements exceptionally well.[31,70,71] Although other high-silica zeolites, e.g. Nu-1 and FU-1[18] are also much more effective catalyst components for xylene isomerization than the silica–aluminas used earlier, ZSM-5-type catalysts are now used in some 90% of US and 60% of existing western world capacity.[31]

The high selectivity of ZSM-5 catalysts for xylene isomerization is due to transition state selectivity (section 4.2). Ethylbenzene is converted over ZSM-5 at low temperatures via transalkylation and at higher temperatures via both trans-alkylation and dealkylation.

In the toluene disproportionation process toluene reacts to give benzene plus xylene. A different selectivity is required here, for the product has to be pre-dominantly para-xylene for an economic process. With ZSM-5 catalysts, the formation of highly substituted aromatics is inhibited as in the xylene isomer-ization process. Selectivity to p-xylene is achieved in a different way, by control of the transport of the products and inhibition of secondary isomerization of the primary p-xylene product.[31,70,71] Various ways of modifying ZSM-5 catalyst to give para-selectivity have been described, these include the use of large crystal sizes, treatment of the zeolite with phosphorus, boron, magnesium, etc. and with coke.[72] Para-selectivity is found to be a function of diffusion properties of the modified ZSM-5 catalysts, independent of the way in which these diffusion properties are controlled.[71] Similar considerations apply to an alternative route for converting toluene selectively to p-xylene, that of methylation with methanol.[31,72] The conversion of methanol itself to hydrocarbons also has to be avoided. Zeolite Nu-10, which probably has 'tight' 10-ring windows, gives selec-tivity to p-xylene in these processes without catalyst modifications.[73]

The vapour phase alkylation of benzene with ethylene to give ethylbenzene was one of the first processes using ZSM-5 catalysts to reach commercial use.[74] Ethylbenzene, the feedstock for styrene, had been made by a process using Friedel-Craft catalysts (usually $AlCl_3$). The new process (Fig. 3.15) developed jointly by Mobil and Badger is cheaper to build and operate and is also free of the corrosion and pollution problems associated with $AlCl_3$. Ten plants using the technology have been announced, corresponding to a total styrene capacity of 3.1×10^6 tonne per year.[74] Although acidic zeolites had been recognized in the early 1960s as active catalysts for the ethylation of benzene, coke formation was too rapid in all the large-pore zeolites then available. With the medium-pore ZSM-5 catalyst, coking is greatly reduced and regeneration cycles of 2–4 weeks are normal. The catalyst shows an exceptionally high activity: at typical conditions of about 400°C and 20 bar pressure, a weight hourly space velocity of up to 400 kg feed kg^{-1} catalyst hour can be used.[70]

The ethylation of toluene to given p-ethyl toluene selectivity is a much more recent process[75] which combines the technology of ethylbenzene production with the para-directing selectivity of modified ZSM-5 catalysts as used in toluene disproportionation. Dehydrogenation gives p-methylstyrene, a monomer giving polymers which have superior properties to polystyrene.

Fig. 3.15. Ethylbenzene/styrene plant with capacity of 700 000 tonnes/year, Borg-Warner Corp. at Carville, Louisiana. The Mobil–Badger technology, including ethylbenzene synthesis over a zeolite ZSM-5 catalyst, is used. (Photograph by courtesy of Badger Co.)

Table 3.11. Alkylation of toluene with ethylene. Composition of typical product steams[75]

Compound	Catalyst		
	HCl-AlCl$_3$	Unmodified ZSM-5	Modified ZSM-5
Light gas/benzene	0.2	1.0	0.9
Toluene (excess recycle)	48.3	74.4	86.2
C$_8$ Aromatics	1.2	1.2	0.5
p-Ethyltoluene	11.9	7.0	11.9
m-Ethyltoluene	19.3	7.0	0.4
o-Ethyltoluene	3.8	0.3	0
C$_{10}^+$ Aromatics	14.4	1.4	0.1
Tar	0.9	0	0
Ethyltoluene isomers (%)			
Para	34.0	31.8	96.7
Meta	55.1	66.8	3.3
Ortho	10.9	1.4	0

6.4   Olefins conversion processes

Three stages of reaction can be recognized[70,76-78] in olefin conversion processes over high-silica zeolites (ignoring olefin isomerization, which is fast even at low temperatures).[79]

1 At low temperatures ($< 150°C$) simple oligomerization occurs, giving largely linear oligomers due to the constraints of the pore system, e.g. with propene.

$$C_3H_6 \rightarrow C_6H_{12} \rightarrow C_9H_{18} \rightarrow C_{12}H_{24}. \ldots$$

2 In the middle temperature range (about 150–300°C) cracking of the oligomers becomes significant, so leading to a redistribution of olefins. All carbon numbers are formed, independent of the starting olefin. Again, from propene,

$$C_3H_6 + C_6H_{12} \rightleftharpoons C_9H_{18} \rightleftharpoons C_4H_{18} + C_5C_{10}$$
$$C_3H_6 + C_4H_8 \rightleftharpoons C_7H_{14}$$
$$C_4H_8 + C_4H_8 \rightleftharpoons C_8H_{16} \rightleftharpoons C_3H_6 + C_5H_{10}$$

3 In the high temperature range ($> 300°C$), hydrogen transfer reactions give overall loss of olefins to aromatics and paraffins. An example of formal reaction from a mixture of olefins (formed as in stage 2) is:

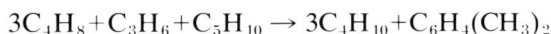

$$3C_4H_8 + C_3H_6 + C_5H_{10} \rightarrow 3C_4H_{10} + C_6H_4(CH_3)_2$$

Mobil have recently announced[80] a process based on stages 1 and 2. Called MOGD (Mobile olefins to gasoline and distillate), it uses a ZSM-5 catalyst to convert light olefins to oligomers which are predominantly straight chain with occasional methyl branches. Variations of process conditions gives either gasoline or distillate as the main product (Table 3.12). The distillate is hydrotreated to give jet or diesel fuel. Both gasoline and diesel fuel are of high quality.

Another new process, Cyclar, has been jointly developed by BP and UOP.[81] This uses all three stages of olefin reactions in the conversion of LPG to aromatics.

Table 3.12. Process yields from a propene/butene feed to MOGD process[80]

Product	Type of process operation	
	Maximum distillate mode	Gasoline mode
C_1–C_3	1	4
C_4	2	5
C_5— 165°C gasoline	18	—
165°C + distillate	79	—
C_5— 200°C gasoline	—	84
200°C + distillate	—	7

Dehydrogenation of the paraffinic components of LPG must be a preliminary step. The process design is based on UOP's Continuous Platforming Process and the catalyst is believed to be a gallium-doped ZSM-5.

6.5   Methanol conversion processes

The conversion of methanol to gasoline over a ZSM-5 catalyst is a totally new process;[37,82] indeed, in combination with the methanol synthesis process, it provided the first new process for the conversion of synthesis gas to liquid hydrocarbons since the discovery of the Fischer–Tropsch process in the 1920s. Another process for converting synthesis gas to gasoline and diesel fuel over a zeolite catalyst has recently been announced by Union Carbide[83] but no details are yet available. A comparison of the (methanol synthesis and MTG) route with Fischer–Tropsch process is given in Table 3.13, where it can be seen that the new process for gasoline is better in many respects.

Table 3.13. Comparison of ICI LP methanol process + Mobil MTG process with Fischer–Tropsch process for conversion of synthesis gas to gasoline

LP methanol + MTG processes	Fischer–Tropsch process
1 High activity, both catalysts.	1 Low activity catalyst.
2 High selectivity, both catalysts.	2 Low selectivity catalysts.
(a) No CH_4 formed by methanol synthesis catalyst; very low CH_4 formation by MTG catalyst.	(a) Significant CH_4 formation.
(b) Hydrocarbon product from MTG catalyst limited $\sim C_{10}$, i.e. no hydrocarbons of higher bp than gasoline.	(b) High boiling hydrocarbons formed as well as gasoline.
(c) Extensive aromatics formation, i.e. gasoline has high octane number.	(c) No aromatics formed; gasoline fraction needs further processing to get acceptable octane number.
3 Long life, both catalysts. *In situ* regeneration of MTG catalyst.	3 Life adequate for commercial use. Catalyst composition changes during life.

Other zeolites and indeed other acidic catalysts also form hydrocarbons from methanol[37] but none as yet combine the virtues of ZSM-5: high activity, high selectivity and relative freedom from coking. Large-pore zeolites (e.g. Y, mordenite) give higher boiling products and coke readily; zeolite Nu-1 gives[84] the same product range as ZSM-5 but the activity disappears in a few minutes due to rapid coking; the clay montmorillonite initially produces aromatics with a maximum at C_{11} but it is deactivated in less than 30 min by coke.[85]

The overall course of the reaction of methanol was shown[52,82] to be:

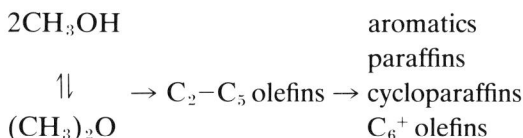

$$
2CH_3OH \quad\updownarrow\quad (CH_3)_2O \quad\rightarrow\quad C_2-C_5 \text{ olefins} \quad\rightarrow\quad
\begin{array}{l}
\text{aromatics} \\
\text{paraffins} \\
\text{cycloparaffins} \\
C_6{}^+ \text{ olefins}
\end{array}
$$

Clearly the second stage of the process—the conversion of light olefins to high-octane gasoline—is essentially as discussed above, but there is much controversy on the mechanism of the conversion of the methanol/dimethyl ether mixture to light olefins. The formation of a trimethyloxonium ion, $(CH_3)_3O^+$, or its ylid form, $(CH_3)_2\overset{+}{O}\overset{-}{C}H_2$, followed by Stevens rearrangement to methyl ethyl ester, or by reaction with methanol, are the more plausible routes.[86] Many other organic materials, e.g. alcohols, esters, plant products such as corn oil, can also be converted to gasoline over ZSM-5 catalysts.[52]

Mobil have developed both fixed-bed and fluid-bed versions of the MTG process.[80,87] The first full scale plant, of 570 000 tonne gasoline year^{-1} due to start up in 1985 in New Zealand, uses the fixed-bed version, mainly because of the simpler engineering.[88] The fluid-bed version has been tested successfully in 4 bbl day^{-1} and 100 bbl day^{-1} pilot plants.[89] The performance of the fluid-bed plant is superior to that of the fixed-bed plant in both gasoline yield and quality.

The conversion of olefins to aromatics can be inhibited to give a methanol to olefins process.[31] Phosphorus-treated ZSM-5 can have up to 70% selectivity to C_2-C_4 olefins.[90] The relative importance of modified acidic sites and modified pore systems is not clear. Smaller-pore zeolites such as FU-1 and Nu-3 also give high yields of C_2-C_4 olefins[18,91] but rapid coking prevents their process use. No large-scale operation has yet been reported.

More complex reactions occur in the liquid phase conversion of methanol over HZSM-5 catalysts, e.g. reaction in the presence of toluene or p-xylene gives substantial amounts of various ketones, carboxylic acid and methyl esters.[92]

7   Future prospects

Catalysis by zeolites, only some 20 years old, is still a bustling subject, likely to produce many more surprises before it reaches maturity. It forms one of the few areas in heterogeneous catalysis where radically new processes and new catalysts of outstanding performance may be found. Advances can be expected on at least four fronts.

(a)   Many more new synthetic zeolites can be expected for the rate of discovery shows no sign of slackening. Each new zeolite is comparable with the availability of a new solvent or reagent for organic chemistry. Possibilities of significance for catalysis include very large pore zeolites (> 12-ring windows), high silica zeolites with 12-ring pores and zeolites (small and large pore) which show as low a rate of coking as ZSM-5.

(b)   The discovery[43] of framework structures analogous to zeolites based on Al_2O_3/P_2O_5 instead of SiO_2/Al_2O_3 opens the way to zeolites with new chemical components as well as new structures.

(c)   So far the process use of zeolite catalysts has depended upon their intrinsic acidity, with or without an added hydrogenation/dehydrogenation component. The use of zeolites as selective 'rigid solvent' containing other catalytic components is at an early stage.[34]

(d)   All the important uses of zeolite catalysts so far have been in the oil refining or petrochemical areas, i.e. with hydrocarbons as feeds or products. The extension of zeolite catalysis to other areas of chemistry already attracts much interest[20,27,34] and novel processes can be expected. Perhaps a pertinent indication is the formation[93] of several amino acids from carbon monoxide and ammonia over X and Y zeolites exchanged with Ca^{2+} and Fe^{3+}.

8   Acknowledgements

I am grateful for the help I have received from Dr W.O. Haag, Dr J.A. Rabo, Dr C. Roberts and Professor J.M. Thomas, from Dr T.V. Whittam and other colleagues in ICI PLC, and from staff in The Badger Co., Laporte Industries Ltd, M.W. Kellogg Co. and Union Carbide Corp.

9   References

General zeolite (including some catalysis)

1   *Proceeedings of the International Conferences on Zeolites* (various publishers). These were held in 1967 (London); 1970 (Worcester, Mass.); 1973 (Zurich); 1977 (Chicago); 1980 (Naples); 1983 (Reno, Nevada).
2   Riekart, L. Sorption, diffusion and catalytic reaction in zeolites. *Adv. Catal.* 1970, **21**, 281.
3   Breck, D.W. *Zeolite Molecular Sieves: Structure, Chemistry and Use.* John Wiley, New York, 1974.
4   Breck, D.W. Synthetic zeolites: properties and applications. In *Industrial Minerals and Rocks* (4th edn) 1975.
5   Schochow F. & Puppe, L. Zeolites—their synthesis, structure and applications. *Angew. Chem. Internat. Edit.* 1975, **14**, 620.
6   Rabo, J.A. & Kasai, P.H., Caging and electrolytic phenomena in zeolites. In *Progress in Solid State Chemistry*, J.O. McCaldin & G. Somorjai (Eds), Vol. 9, p. 1. Pergamon, Oxford, 1975.
7   Barrer, R.M. *Zeolites and Clay Minerals as Sorbates and Molecular Sieves.* Academic Press, London, 1978.
8   Haynes, H.W. Chemical, physical and catalytic properties of large pore acidic zeolites. *Adv. Catal.* 1978, **17**, 273.
9   Uytterhoeven, J.B. Zeolites and their role in sorption and catalysis. *Prog. Colloid and Polymer Sci.* 1978, **65**, 233.
10 Townsend, R.P. (Ed.) *Properties and Applications of Zeolites.* Special Publication No 33. Chemical Society, London, 1980.

11   Flanigen, E.M. Molecular sieve zeolite technology: the first twenty-five years. *Pure & Appl. Chem.* 1980, **52**, 2191.

12   Jacobs, P.A. Acid zeolites: an attempt to develop unifying concepts. *Catal. Rev. Sci. Eng.* 1982, **24**, 415.

13   Barrer, R.M. *Hydrothermal Chemsitry of Zeolites.* Academic Press, London, 1982.

14   Stucky, G.D. & Dwyer, F.G. *Intrazeolite Chemistry,* ACS Symposium Series, Vol. 218, American Chemical Society, Washington D.C. 1983.

15   Fyfe, C.A., Thomas, J.M., Klinowski, J. & Gobbi, G.C. Magic-angle-spinning NMR (MAS-NMR) spectroscopy and the structure of zeolites. *Angew. Chem. Int. Ed.* 1983, **22**, 259.

16   *Zeolites: Science and Technology.* NATO Asi Ser., Ser. E., 1984, **80**.

17   Thomas, J.M. *New Approaches to the Structural Elucidation of Zeolites and Related Catalysts.* Proc 8th Int. Congress Catal. Vol. 1, p. 31. Verlag Chemie, Weinheim. 1984.

18   Dewing, J., Spencer, M.S. & Whittam, T.V. Synthesis, characterisation and catalytic properties of Nu-1, FU-1 and related zeolites. *Catal. Rev. Sci. Eng.* (in press).

19   Jacobs, P.A. *High Silica Zeolites.* Elsevier, Amsterdam (in prep.).

General zeolite catalysis

20   Venuto, P.B. & Landis, P.S. Organic catalysis over crystalline alluminosilicates. *Adv. Catal.* 1968, **18**, 259.

21   Turkevich, J. Zeolites as catalysts. *Catal. Rev.* 1968, **1**, 1.

22   Leach, H.F. Application of molecular sieve zeolites to catalysis. In *Annual Reports, Vol. 68A,* p. 195. Chemical Society, London, 1971.

23   Weisz, P.B. Zeolites—new horizons in catalysis. *Chem. Tech.* 1973, 498.

24   Rabo, J.A. (Ed.) *Zeolite Chemistry and Catalysis.* ACS Monograph 171. American Chemical Society, Washington D.C., 1976.

25   Rudham, R. & Stockwell, A. Catalysis on faujasitic zeolites. In *Catalysis, Specialist Periodical Report,* Kemball, C. (Ed.), Vol. 1, p. 87. Chemical Society, London, 1977.

26   Jacobs, P.A. *Carbiogenic Activity of Zeolites.* Elsevier, Amsterdam, 1977.

27   Venuto, P.B. Aromatic reactions over molecular sieve catalysts: a mechanistic review. In *Catalysis in Organic Synthesis* 1977, G.V. Smith (Ed.), p. 67. Academic Press, New York, 1977.

28   Venuto, P.B. & Habib, E.T. Catalyst–feedstock–engineering interactions in fluid catalytic cracking. *Catal. Rev. Sci. Eng.* 1978, **18**, 1. *Fluid Catalytic Cracking with Zeolite Catalysts.* Marcel Dekker, New York, 1979.

29   Naccache, C. & Taarit, Y.B. Recent developments in catalysis by zeolites. In *Proc. 5th Int. Conf. Zeolites,* L.V.C. Rees (ed.), p. 592. Heyden, London, 1980.

30   Spencer, M.S. & Whittam, T.V. Catalysis on non-faujasitic zeolites and other strongly acidic oxides. In *Catalysis, Specialist Periodical Report,* Kemball, C., & Dowden, D.A. (Eds), Vol. 3, p. 189. Chemical Society, London, 1980.

31   Weisz, P.B. *Molecular Shape Selective Catalysis,* 7th International Congress on Catalysis, Tokyo, 1980.

32   Imelik, B. *et al.* (Eds.) *Catalysis by Zeolites,* Studies in Surface Science and Catalysis No. 5. Elsevier, Amsterdam, 1980.

33   Kerr, G.T., Rabo, J.A. & Heinemann, H. Zeolite catalysis. *Catal. Rev. Sci. Eng.* 1981, **23**, 281, 293, 315.

34   Maxwell, I.E. Nonacid catalysis with zeolites. *Adv. Catal.* 1982, **31**, 1.

35   Whyte, T.E. & Dalla Betta, R.A., Zeolite advances in the chemical and fuel industries: a technical perspective. *Catal. Rev. Sci. Eng.* 1982, **24**, 567.

36   Selectivity in heterogeneous catalysis. *Faraday Disc. Chem. Soc.* 1981, **72**.

37   Chang, C.D. Hydrocarbons from methanol. *Catal. Rev. Sci. Eng.* 1983, **25**, 1.

38   Minachev, K.M. & Kondratev, D.A. Properties and use in catalysis of zeolties of the pentasil type. *Russ. Chem. Rev.* 1983, **52**, 1113.

39 *Advances in Catalytic Cracking.* Preprints, Amer. Chem. Soc., Divn, Petroleum Chem., August 1983, **28** (4), 861–952.
40 Csicsery, S.M. Shape-selective catalysis in zeolites. *Zeolites* 1984, **4**, 202.

Other references
41 Wells, A.F. *Structural Inorganic Chemistry*, 4th edn, p. 803. Clarendon Press, Oxford, 1975.
42 Haag, W.O., Lago, R.M. & Weisz, R.P. *Nature* 14 June 1984, **309**, 589.
43 Wilson, S.T., Lok, B.M., Messina, C.A., Cannon, T.R. & Flanigen, E.M. *J. Amer. Chem. Soc.* 1982, **104**, 1145. Bennet, J.M., Cohen, J.P., Flanigen, E.M., Pluth, J.J. & Smith, J.V. Ref. 14, p. 109. Wilson, S.T., Lok, B.M., Messina, C.A., Cannon, T.R., & Flanigen, E.M. Ref. 14, p. 70.
44 Thomas, J.M. & Millward, G.R. *J. Chem. Soc., Chem. Commun.* 1982, 1380.
45 Olson, D.H., Haag, W.O. & Lago, R.M. *J. Catal.* 1980, **61**, 390; Jacobs, P.A. & von Ballmoos, R. *J. Phys. Chem.* 1980, **86**, 3050.
46 Irving, J.D.N., Leach, H.F., Spencer, M.S. & Whan, D.A. *J. Chem. Res.* (S), 1983, 136; (M), 1983, 1401.
47 Barrer, R.M. *Zeolites* 1981, **1**, 80.
48 Poutsma, M.L. Ref. 24, p. 437.
49 Namba, S., Nakanishi, S. & Yashima, T. *J. Catal.* 1984, **88**, 505; Gilson, J.P. & Derouane, E.G. *J. Catal.* 1984, **88**, 538.
50 Csicsery, S.M. Ref. 24, p. 680.
51 Haag, W.O., Lago, R.M. & Weisz, P.B. *Faraday Disc. Chem. Soc.* 1982, **72**, 317.
52 Haag, W.O., Olson, D.H. & Weisz, P.B. In *Chemistry for the Future*, Grunewald, H. (Ed.), p. 327. Pergamon, Oxford, 1984.
53 Olson, D.H. & Haag, W.O. In *Catalytic Materials: Relationship between Structure and Reactivity.* White, T.E., Dalla Betta, R.A., Derouane, E.G. & Baker, R.T.K. (Eds.) Symposium Series, No. 248, p. 275. Amer. Chem Soc., Washington D.C., 1984.
54 Walsh, D.E. & Rollman, L.D. *J. Catal.* 1979, **56**, 195.
55 Knight, W.N.N. & Peniston-Bird, M.L. In *Modern Petroleum Technology*, 4th edn, Hobson, G.D. & Phol, W. (Eds), p. 278. Applied Science, Barking, 1973.
56 Jahnig, C.E., Martin, H.Z. & Campbell, D.L. *Chemtech* 1984, 106.
57 *Oil Gas J.* 26 December 1983, **81** (52), 77.
58 *Oil Gas J.* 19 March 1984, **82** (12), 84.
59 Haag, W.O. & Dessau, R.M. *Proc. 8th Int. Congress Catal.* Vol. 2, p. 305. Verlag Chemie, Weinheim, 1984.
60 Plank, C.J., Rosinski, E.J. & Hawthorne, W.P. *Ind. Eng. Chem., Prod. Res. Dev.* 1964, **3**, 165; Plank, C.J. & Rosinski, E.J. *Chem. Eng. Prog. Symposium Series* No. 73, 1967, **63**, 26; Plank, C.J. *Chemtech*, 1984, 243.
61 Luckenbach, E.C. *Chem. Eng. Prog.* February 1979, **75** (2), 56.
62 McArthur, D.P., Simpson, H.D. & Baron, K. *Oil Gas J.* 23 February 1981, **79** (8), 55; Baron, D., Wu, A.H. & Drenzbe, L.D. Ref. 39, p. 934.
63 Magee, J.S., Ritter, R.E. & Rheaume, I. *Hydrocarbon Process*, September 1979, **58** (9), 123.
64 Tolen, D.F. *Oil Gas J.* 30 March 1981, **79** (13), 90.
65 Reagan, W.J., Wolterman, G.M. & Brown, S.M. Preprints, Amer. Chem. Soc., Divn. Petroleum Chem., August 1983, **28** (4), 884.
66 *Oil Gas J.* 18 June 1984, **82** (25), 86.
67 Ward, J.W. *Hydrocarbon Processing* Sept. 1975, 101. Mavity, V.T., Ward, J.W. & Whitehead, K.E. *Hydrocarbon Processing* Nov. 1978, 157.
68 Chen, N.Y., Goring, R.L., Ireland, H.R. & Stein, T.R. *Oil Gas J.* 6 June 1977, **75** (23), 165. Perry, R.H., Davis, F.E. & Smith, R.B. *Oil Gas J.* 22 May 1978, **76**, (21), 78. Ireland, H.R., Redini, C., Raff, A.S. & Fava, L. *Oil Gas J.* 11 June 1979, **77** (24), 82.
69 Donnelly, S.P. & Green, J.R. *Oil Gas J.* 27 Oct 1980, **78** (43), 77.
70 Haag, W.O., Olsen, D.H. & Weisz, P.B. In *Chemistry for the Future.* Grunewald, H. (Ed.), p. 327. Pergamon Press, Oxford, 1984.
71 Olsen, D.H. & Haag, W.O. In *Catalytic Materials: Relationship between Structure and Reactivity.*

Whyte, T.E., Dalla Betta, R.A., Derouane, E.G. & Baker, R.T.K. (Eds), p. 275. ACS Symposium Series, No. 248, 1984.

72 Chen, N.Y., Kaeding, W.W. & Dwyer, F.G. *J. Amer. Chem. Soc.* 1979, **101**, 6783.
Kaeding, W.W., Chu, C., Young, L.B. & Butter, S.A. *J. Catal.* 1981, **69**, 392.
Kaeding, W.W., Chu, C., Young, L.B. & Winstein, B., *J. Catal.* 1981, **67**, 159.
Young, L.B., Butter, S.A. & Kaeding, W.W. *J. Catal.* 1982, **76**, 418.

73 Araya, A. & Lowe, B.M. *Zeolites* 1984, **4**, 280.
Hogan, P.J., Whittam, T.V., Birtill, J.J. & Stewart, A. *Zeolites* 1984, **4**, 275.

74 Dwyer, F.G., Lewis, P.J. & Schneider, F.H. *Chem. Eng.* 5 Jan 1976, **83** (1), 90.
Lewis, P.J. & Dwyer, F.G. *Oil Gas J.* 26 Sept 1977, **75** (40), 55.
Hagopian, C.R., Lewis, P.J. & MacDonald, J.J. *Hydrocarbon Proc.* Feb 1983, **62** (2), 45; *Hydrocarbon Proc.*, Nov 1983, **62** (11), 148.

75 Kaeding, W.W., Young, L.B. & Prapas, A.G. *Chem. Tech.* 1982, **12**, 556.
Kaeding, W.W. & Bavile, G.C. *New Monomers and Polymers*, Culbertson, B.M. & Pittman, C.V. (Eds), p. 223. Plenum Corp., New York, 1984.

76 Anderson, J.R., Mole, T. & Christov, V. *J. Catal.* 1980, **61**, 477.

77 Dejaifvr, P., Vedrine, J.C., Bolis, V. & Derouane, E.G. *J. Catal.* 1980, **61**, 331.

78 van den Berg, J.P., Wolthuizen, J.P., Clagne, A.D.H., Hays, G.R., Huis, R. & van Hooff, J.H.C. *J. Catal.* 1983, **80**, 130.
van den Berg, J.P., Wolthaizen, J.P. & van Hooff, J.H.C. *J. Catal.* 1983, **80**, 139.

79 Irving, J.D.N., Leach, H.F., Whan, D.A. & Spencer, M.S., *J. Chem. Res.*, (S) 1982, 66, (M) 1982, 0852.

80 Tabak, S.A. Papers at Thailand—US Natural Gas Utilization Symposium, Bangkok, Thailand, 7–11 Feb, 1984; *Nonpetroleum Vehicular Fuels IV*, Arlington, Virginia, USA, 16–18 April 1984.

81 Anon, *Petroleum Rev.*, July 1984, **38** (450), 33.
Johnson, J.A. & Hilder, G.D. *NPRA Annual Meeting*, San Antonio, Texas, March 1984, Paper 45.

82 Meisel, S.P., McCullough, J.P., Lechthaler, C.H. & Weisz, P.B. *Chemtech* Feb 1976, 86.
Chang, C.D. & Silvestri, A.J. *J. Catal.* 1977, **47**, 249.

83 *Hydrocarbon Proc.* Feb 1984, **63** (2), 23.

84 Spencer, M.S. & Whittam, T.V. *Acta Phys. Chem.* 1978, **24**, 307.

85 Meisel, S.L. & Weisz, P.B. *Symposium on Advances in Catalytic Chemistry II*, University of Utah, 18–21 May, 1982.

86 van den Berg, J.P., Walthuizen, J.P. & van Hooff, J.H.C., *Proc. 5th Int. Conf. Zeolites*. Rees, L.V.C. (Ed.), 1980, p. 649.
Mole, T. & Whiteside, J.A. *J. Catal.* 1982, **75**, 284.
Haag, W.O., Lago, R.M. & Rodewald, P.G. *J. Mol. Catal.* 1982, **17**, 161.
Wu, M.M. & Kaeding, W.W. *J. Catal.* 1984, **88**, 478.
Olah, G.A., Doggweiler, H., Felberg, J.D., Frohlich, S., Gordina, M.J., Karpeles, R., Keumi, T., Inaba, S., Ip, W.M., Lammerstsma, K., Salem, G. & Tabor, D.C. *J. Amer. Chem. Soc.* 1984, **106**, 2143.

87 Lee, W., Mazuik, J., Weekman, V.W. & Yurchak, S. *Chem. Eng. Monog.* 1979, **10**, 171.
Chang, C.D., Kuo, J.C.W., Lang, W.H., Jacob, S.M., Wise, J.J. & Silvestri, A.J. *Ind. Eng. Chem.*, *Process Des. Dev.* 1978, **17**, 255.
Yurchak, S., Voltz, S.E. & Warner, J.P. *Ind. Eng. Chem.*, *Process, Des. Dev.* 1979, **18**, 527.

88 *Oil Gas J.* 14 Jan 1980, **78** (2), 95. *Chem. Eng.* 7 April 1980, 43. *Oil Gas J.* 14 May 1984, **82** (20), 76.

89 Liederman, D., Jacob, S.M., Voltz, S.E. & Wise, J.J. *Ind. Eng. Chem.*, *Process Des. Dev.* 1978, **17**, 340. *Chem. Week*, 25 Jan 1978, 35. *Oil Gas J.* 15 Aug. 1983, 31.

90 Kaeding, W.W. & Butter, S.A. *J. Catal.* 1980, **61**, 155.

91 Spencer, M.S. & Whittam, T.V. *J. Mol. Catal.* 1982, **17**, 271.

92 Deane, S., Wilshiev, K., Western, R., Mole, T. & Seddon, D. *J. Catal.* 1984, **88**, 499.

93 Fripiat, J.J., Poucelet, G., van Assche, A.T. & Magaudon, J. *Clays and Clay Minerals*, 1972, **20**, 331.

4 Catalysis in ammonia and methanol production

J.R. Jennings and M.V. Twigg

1 Introduction

Ammonia and methanol, both large tonnage commodity chemicals, are manu-factured by multistage catalytic processes using common feedstocks. For the production of ammonia from natural gas or naphtha, the sequence of process steps is as follows: hydrodesulphurization, hydrogen sulphide removal, primary and secondary steam reforming, high and low temperature shift, carbon dioxide removal, methanation and ammonia synthesis. For methanol production, the product from the primary reforming step can be fed after water removal to the synthesis reactor though, depending on the hydrocarbon feedstock used, additional carbon dioxide is frequently added. Most of the process steps use heterogeneous catalysts, the two exceptions being hydrogen sulphide and carbon dioxide removal. The types of catalyst commonly used in each stage are listed in Table 4.1, which also includes approximate data on tonnages and lifetimes typical of modern plants. Ammonia and methanol are often manufactured together on the same site, where depending upon market demands, surplus hydrogen from methanol manufacture can be fed to the ammonia process, or alternatively surplus carbon dioxide from ammonia manufacture can be fed to the methanol loop. Such is the case in the ICI complex at Billingham, England. A simplified flow diagram is shown in Fig. 4.1; it should be stressed that ammonia and methanol units have their own reforming capacity.

Table 4.1. Catalysts used in the production of ammonia and methanol from hydrocarbons

Catalyst	Composition	Physical form*	Quantity (tonnes)‡	Life‡ (years)
Hydrodesulphurization	Co + Mo sulphides on alumina	Extrudate or granules	15	7–10
Hydrogen sulphide removal	ZnO + binders	Granules	25	2–5
Primary steam reforming	Ni on ceramic support	Rings	20	2–5
Secondary steam reforming	Ni on ceramic support	Rings	30	6–10
High temperature shift	Fe_3O_4/Cr_2O_3	Pellets	55	2–4
Low temperature shift	$Cu/ZnO/Al_2O_3$	Pellets	60	2–4
Methanation	Ni/Al_2O_3 + binders	Pellets	30	5–10
Ammonia synthesis	$Fe/Al_2O_3/CaO/K_2O$	Granules	170§	5–10
Methanol synthesis	$Cu/ZnO/Al_2O_3$	Pellets	120§	3–4

*Often a particular catalyst may be available in several shapes and sizes. The one used depends on factors such as reactor design, operating conditions and cost considerations.
‡Typical quantities used in a 1000 tonnes day^{-1} plant.
‡Typical lives depend on the duty, kind of plant and catalyst used.
§The amount of synthesis catalyst used in a particular plant largely depends on the loop pressure.

The world market for ammonia is some 90×10^6 tonnes year^{-1} with an hydroapproximate growth rate of 2%. The world market for methanol is 12×10^6 tonnes year^{-1}, but in this case the future growth rate is much less certain and depends largely upon the extent to which methanol will be used as a convenient liquid fuel, independent of crude oil.

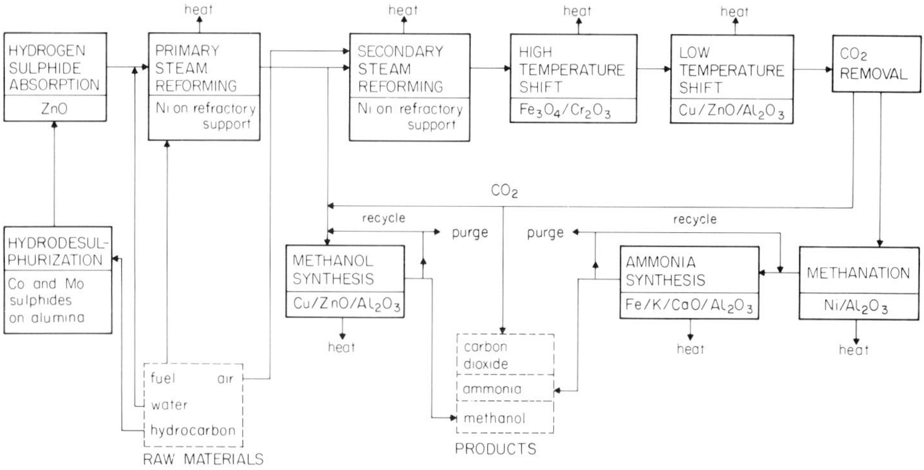

heat   heat   heat   heat

HYDROGEN SULPHIDE ABSORPTION	PRIMARY STEAM REFORMING	SECONDARY STEAM REFORMING	HIGH TEMPERATURE SHIFT	LOW TEMPERATURE SHIFT	CO_2 REMOVAL
ZnO	Ni on refractory support	Ni on refractory support	Fe_3O_4/Cr_2O_3	$Cu/ZnO/Al_2O_3$	

CO_2

recycle   purge   purge   recycle

HYDRODESUL-PHURIZATION	METHANOL SYNTHESIS		AMMONIA SYNTHESIS	METHANATION
Co and Mo sulphides on alumina	$Cu/ZnO/Al_2O_3$		$Fe/K/CaO/Al_2O_3$	Ni/Al_2O_3

heat   carbon dioxide   heat   heat

fuel   air   ammonia

water

hydrocarbon   methanol

RAW MATERIALS   PRODUCTS

Fig. 4.1. Schematic diagram illustrating the production of ammonia and methanol from hydrocarbon feedstock.

This review, concerned mainly with catalysis, gives a brief history of ammonia and methanol catalysis and then goes on to describe production of the synthesis gases for both products. The synthesis reactions are then discussed in some detail followed by brief conclusions.

2   Historical aspects

2.1   Ammonia

Towards the end of the nineteenth century, the demand for nitrogenous fertilizer was beginning to exceed supply, which came mostly from Chile in the form of sodium nitrate. Some ammonia was available as a by-product from the destructive distillation of coal and a little more came from hydrolysis of calcium cyanamide. The generation of ammonia from atmospheric nitrogen to meet growing demands for fertilizers and explosives was an international priority. The problem of the inertness of nitrogen towards chemical attack was solved largely through the efforts of Fritz Haber, Carl Bosch, Alwin Mittasch and co-workers who, over a period of a decade, developed a laboratory observation into a 30 tonnes day^{-1} commercial catalytic process using radically new technology. Never before had such pressures and temperature been used in the chemical industry. With the onset of World War I, the need for explosives resulted in extensive additions to synthetic ammonia capacity. This was achieved initially by a tenfold increase in cyanamide production, but the main contribution came from 'catalytic ammonia'. From 6500 tonnes year^{-1} in 1913, production increased to 200000 tonnes year^{-1}

Fig. 4.2. The ammonia synthesis converter of the ICI plc Ammonia IV plant at Billingham, UK. (By courtesy of ICI plc.)

by the end of World War I. The scale of operation has grown significantly since the early days. Large single stream plants based on hydrocarbon steam reforming established in the mid 1960s typically produced up to 1000 tonnes day^{-1} while the ICI plant built at Billingham in 1976 (Fig. 4.2), has a capacity in excess of 1300 tonnes day^{-1}, some 35 times greater than the original BASF plant: even larger plants have been built since then.

The first indication that a catalytic process for the production of ammonia from nitrogen and hydrogen might be feasible was found in 1884 when Ramsay & Young[1] demonstrated that decomposition of ammonia to nitrogen and hydrogen over iron catalysts at 800°C did not go to completion and an equilibrium ammonia level was always obtained. Their attempts to reverse the reaction were unsuccessful except when moist gases were used. Traces of ammonia were obtained by Perman & Atkinson[2] in 1904, over iron catalysts at atmospheric pressure and later in 1907, Nernst et al.[3] reported that considerable quantities of ammonia were obtained using platinum, manganese or iron catalysts at 75 bar. Haber had been working on the nitrogen/hydrogen/ammonia equilibrium at atmospheric pressure and with Le Rossignol began work at elevated pressure, which culminated in 1909 with a demonstration to BASF of a laboratory-scale process for the production of ammonia from its elements over uranium or osmium catalysts at 550°C and 200 bar.[4]

The development of this process to an industrial scale was undertaken by Bosch and Mittasch at BASF where in 1910 the potash-promoted iron catalyst was discovered.[5] Progress followed rapidly and in 1913, the first commercial 'Haber–Bosch Process' went on stream.[6] The levels of technical achievement, both scientific and engineering, were recognized by the award of Nobel Prizes[7] to Haber in 1919 and Bosch in 1931.

2.2   Methanol

Methanol, discovered in 1661 by Robert Boyle, was manufactured until early in the twentieth century by the destructive distillation of wood. Studies in the high pressure technology previously developed for ammonia manufacture turned to carbon monoxide hydrogenation, and it is no surprise that the first synthetic methanol plant was commissioned by BASF in 1923. The process operated at ~ 400°C, and ~ 200 bar, at a scale of 10–20 tonnes day^{-1} using a zinc oxide/chromium oxide catalyst. The basic process operated until the mid 1960s, by which time the scale of operation had reached about 450 tonnes day^{-1}.

Hydrogenation of carbon monoxide was a subject of active research in the early years of the twentienth century. Sabatier[8] had discovered the methanation reaction (hydrogenation of carbon monoxide to methane) using nickel catalysts in 1901 and later work in BASF showed that mixtures of alcohols and hydrocarbons could be obtained using cobalt, iron, osmium or zinc-based catalysts. This work formed the basis from which the Fischer–Tropsch process for the production of liquid hydrocarbons from coal was later developed.[9] The breakthrough for methanol production came in 1923 with the work of Mittasch, Pier & Winkler[10] who discovered that zinc oxide–chromium oxide catalysts were highly selective for the direct synthesis of methanol. These and related catalysts were used in the initial high pressure methanol plants, some of which have survived to this day.

In common with ammonia, once a methanol synthesis catalyst had been developed most of the improvements in the process were around the generation of synthesis gas. Catalysts based on copper/zinc oxide/alumina and operating at 150–250 bar were introduced in Poland in 1952[11] and in 1966 the 'high pressure' technology was superseded by the introduction of the ICI low pressure process, using similar catalysts. The scale of production increased rapidly using this technology and plants with a capacity of around 1500 tonnes day^{-1} are now commonplace. A general view of a typical methanol plant is shown in Fig. 4.3.

Fig. 4.3. General view of the Methanol 2 plant at Billingham, UK. (By courtesy of ICI plc.)

2.3   Hydrogen generation

With the successful development of active synthesis catalysts, the similarities between an ammonia and a methanol plant become more apparent. The key to both processes is the economic production of hydrogen. The original process operated by BASF was the water gas process in which steam reacts with coke at red heat. This produces a mixture of hydrogen, carbon monoxide and carbon dioxide according to equations (1) and (2). For use in the methanol process, the hydrogen to carbon monoxide ratio was adjusted to 2:1 by the shift reaction, equation (3), and carbon dioxide was removed by dissolution in water under pressure. Poisons such as iron carbonyl and sulphur compounds were removed and the purified synthesis gas passed to the converter.

$$C + H_2O \rightarrow CO + H_2 \qquad \Delta H_{298} = +131.2 \, kJ \, mol^{-1} \tag{1}$$

$$C + 2H_2O \rightarrow CO_2 + 2H_2 \qquad \Delta H_{298} = +90.1 \, kJ \, mol^{-1} \tag{2}$$

$$CO + H_2O \rightarrow CO_2 + H_2 \qquad \Delta H_{298} = -41.1 \, kJ \, mol^{-1} \tag{3}$$

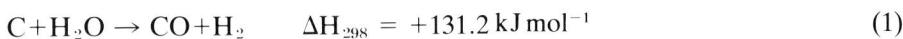

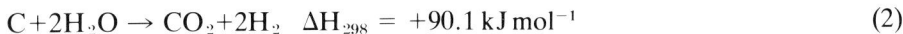

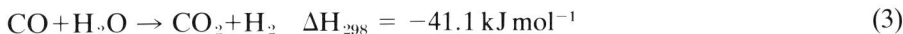

The case of ammonia is more complex in that nitrogen needs to be introduced into the process stream and that all carbon oxides must be removed prior to synthesis. The heat for the water gas reaction was provided by blowing air through the hot coke. Sufficient product from this strongly exothermic combustion process containing mainly carbon monoxide and nitrogen, so called producer gas, was added to the water gas to produce an ultimate hydrogen nitrogen ratio of 3:1. The carbon monoxide level was reduced to about 4% by the water gas shift reaction at 300–500°C using iron catalysts containing Fe_3O_4 as the active phase, which also forms the basis of the high temperature shift catalyst in use today. Carbon dioxide was removed in a similar manner to the methanol process. The residual carbon monoxide was removed by absorption at pressure in an ammoniacal cuprous salt solution. The purified nitrogen/hydrogen mixture thus prepared was then compressed and fed to the synthesis reactor.

Coal remained the main raw material for ammonia production in Europe until after World War II, though steam reforming of natural gas was developed in the US from about 1930. Partial oxidation of hydrocarbons was introduced briefly in the 1950s, but the major breakthrough in Europe was the development of nickel-based catalysts for naphtha steam reforming at pressure, by ICI. At that time in Europe there was a plentiful supply of cheap naphtha, whose use was significantly more economic than that of coal and which not only eliminated the cumbersome solids handling procedures associated with the use of coal, but also produced much purer gas streams (lower sulphur content) and dramatically improved the environment in the vicinity of the plant. A consequence of the improved quality of the process gas was that catalysts based on copper and zinc oxide could be used. Though copper-based catalysts had previously been recognized as active for both the shift reaction and methanol synthesis, they needed clean gas streams before commercialization could be realized. Thus, low sulphur levels arising through this change of feedstock enabled development of highly active copper-based low temperature shift and the ICI low pressure methanol catalysts. Though methanation using iron catalysts was used in some old ammonia plants, a further development in gas purification was the use of nickel-based methanation catalysts, derived from Sabatier's early work, which finally eliminated the ammoniacal cuprous liquor absorption process.

Modern ammonia and methanol plants are large single-stream units, which may be integrated so that hydrogen and carbon dioxide may be transferred from one to another depending upon the needs at a given time. With modern synthesis catalysts, the key to successful commercial operation lies in the effective production of synthesis gas, and catalysis still has much to offer in the pursuit of this goal.

3   Synthesis gas: production and purification

3.1   General considerations

Synthesis gas is the term usually applied to a mixture of hydrogen and carbon oxides.* It is an important intermediate in the chemical industry that, with the aid of appropriate catalysts, is used in the manufacture of a wide variety of chemicals that include ammonia, methanol, detergents and acetic acid. Recently, synthesis gas has been used for the reduction of iron ore and this application is likely to consume very large quantities of synthesis gas in the future. At the moment ammonia production consumes the largest quantity of synthesis gas ($\sim 90 \times 10^6$ tonnes of ammonia year^{-1}), with that for methanol being second in importance.

Although coal was the original feedstock for the production of synthesis gas, almost all synthesis gas is now derived from hydrocarbons, and some 10^8 tonnes of hydrocarbon are converted into synthesis gas each year. Two processes are commonly used: 'catalytic steam reforming' and 'partial oxidation' and both make use of the strongly endothermic reaction with water typified by the methane-steam reaction of equation (4). In both processes the exothermic water gas shift reaction in equation (3) is also brought to equilibrium, so hydrogen, carbon monoxide and carbon dioxide are formed.

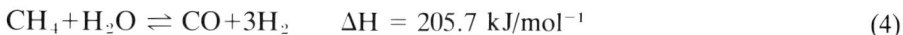

$$CH_4 + H_2O \rightleftharpoons CO + 3H_2 \qquad \Delta H = 205.7 \ kJ/mol^{-1} \qquad (4)$$

At temperatures below $\sim 500°C$ the equilibrium constant for the hydrocarbon steam reaction is very small and only becomes favourable at $\sim 700°C$ and above (see Fig. 4.4). However, the *rate* of the homogeneous gas phase reaction is slow below $\sim 1000°C$ relative to methane cracking. Carbon removal via the water–gas reaction is also slow at these temperatures and deposition of carbon is significant.

3.2   Partial oxidation of hydrocarbons

A fundamental difference between partial oxidation and catalytic steam reforming is the way in which heat is generated for the reaction. In partial oxidation the heat is generated by *in situ* combustion of the feedstock with a deficiency of oxygen. Typically, reaction conditions for non-catalytic partial oxidations are $\sim 1350°C$ at 30–80 bar, and several processes have been developed that are in operation around the world.

Advantages of this approach include the possibility of using a range of hydrocarbons as feedstock (including less desirable fractions) and purity is not critical.

*In the ammonia industry the approximate 3:1 mixture of hydrogen and nitrogen is also called synthesis gas.

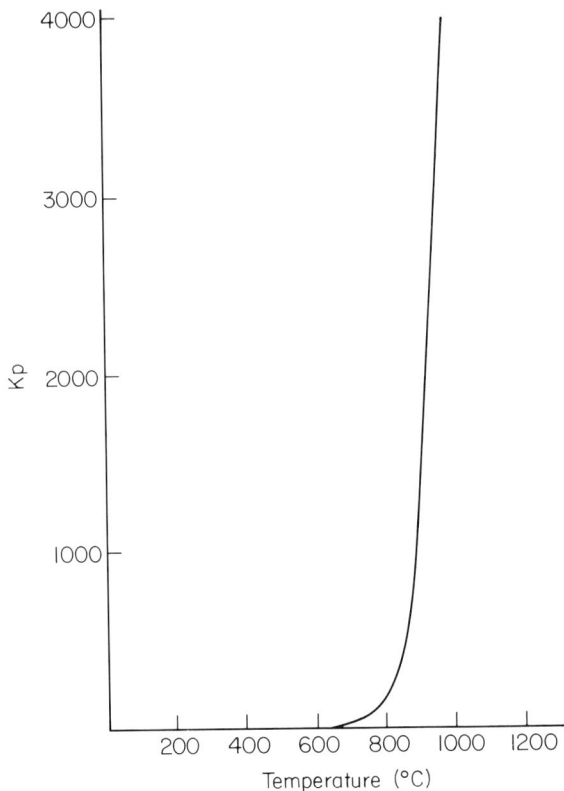

Fig. 4.4. Variation of the equilibrium constant with temperature for the reaction of methane with water to form carbon monoxide and hydrogen.

An interesting development is the possibility of using a catalytic process, in which feedstock purity becomes more important. However, partial oxidation has the disadvantage of requiring an air separation plant to supply oxygen (or enriched air) and in practice large quantities of soot are formed that can cause problems. Overall, the present efficiency of partial oxidation routes to synthesis gas is not as high as that of steam reforming. The product gas is relatively carbon-rich with a high carbon monoxide to carbon dioxide ratio resulting from the high temperature of operation (equation (3) is exothermic so the equilibrium lies to the left at high temperatures and the low steam ratio also favours carbon monoxide formation). Both of these factors are undesirable for methanol and ammonia manufacture; thus, catalytic steam reforming is normally used for producing synthesis gas for these duties.

3.3   Catalytic steam reforming

3.3.1   General. The presence of a suitable catalyst enables the hydrocarbon/ steam equilibrium to be established at a lower temperature than is used in partial oxidation. Heat to drive the endothermic reaction is generated external to the process stream in catalytic steam reforming, and because of the large heat loads involved in a modern plant it is necessary to use tubular reactors. The steam reforming of naphtha displaced the production of synthesis gas from coal in the early 1960s, but this in turn was replaced by natural gas steam reforming following the discovery of gas reserves in the North Sea during the 1970s.

3.3.2   Steam reforming catalysts. Several metals are catalysts for hydrocarbon steam reforming. The precious metals, rhodium and ruthenium are particularly active,[12] while in comparison cobalt and nickel have moderate activity; but as a consequence of their relative cost, nickel catalysts are used industrially.[13] Catalyst in the form of Raschig rings (typically 17 mm in diameter and 10 mm high) containing the active nickel supported on a refractory carrier, is contained in tubes about 10 cm diameter and some 10–15 m long, through which pass the reactant mixtures of steam and hydrocarbon at a pressure of ~ 30 bar. The special spun-cast, high performance alloy tubes have walls typically 20 mm thick and their outer surfaces are heated to temperatures approaching 1000°C by suspending them in a furnace. A modern steam reformer operating at 20–30 bar and producing some 1000 tonnes of synthesis gas day^{-1} may contain 200–400 or even more tubes. Depending on the particular installation the heat load is in the region of 50–100 MW.

Steam reforming catalysts are required to operate under demanding conditions—at temperatures up to perhaps 850°C in the presence of a high partial pressure of steam (~ 20 bar)—and to retain good activity while maintaining their physical properties. Over the years steam reforming catalysts have been improved and now, depending on the duty, have lives up to 5 years though 2–3 years is more common in highly rated plants. The modern catalysts for natural gas reforming are usually made by impregnating a preformed support with a suitable nickel salt, which is subsequently thermally decomposed to the oxide.[14] Depending on the application the finished catalyst may contain between 10 and 30% by weight of nickel oxide. To achieve this level of nickel, several impregnation/calcination cycles may be needed. The porous refractory support must be capable of withstanding the process conditions and materials used include α-alumina, calcium aluminate and magnesia/alumina spinel. The catalyst is most commonly used in the form of Raschig rings since this provides a good compromise between geometric surface area and pressure-drop characteristics, the latter being an important consideration with such long (e.g. 12 m) tubular reactors. Catalyst breakage during use can lead to high pressure-drop across the reformer that might restrict throughput and overall plant efficiency.

3.3.3 *Carbon formation in steam reforming.* There are several reactions that could produce carbon in the steam reforming of hydrocarbons. These are thermal cracking of the hydrocarbon, carbon monoxide disproportionation and the reduction of carbon monoxide (equations (5)–(7)). In conventional steam reformers thermal cracking is the most important route, although carbon monoxide disproportionation is of greater concern at low steam ratios as encountered in processes used to provide gas for the reduction of iron ore.

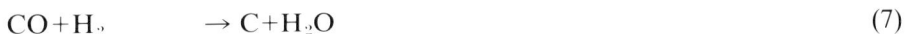

$$CH_3(CH_2)_n CH_3 \rightarrow (2+n)C+(3+n)H_2 \tag{5}$$

$$2CO \rightarrow C+CO_2 \tag{6}$$

$$CO+H_2 \rightarrow C+H_2O \tag{7}$$

The carbon produced in these reactions may be more or less graphitic. Various forms have been observed including amorphous, and long tubular, whisker-like filaments with nickel crystallites at one end. Such forms of carbon have significantly higher Gibbs free energies than graphite (up to 10 kJ mol^{-1}) when they are formed at relatively low temperatures, while deviations from graphite diminish at high temperatures.[15] These thermodynamic effects modify the conditions under which carbon can form, and are taken into account when designing steam reformers.

Carbon formation in natural gas steam reforming. Economic considerations demand that plants usually operate with as low a ratio of steam to hydrocarbon as possible without incurring carbon formation. The presence of carbon in the tube reduces heat transfer into the tube causing the tube wall temperature to rise markedly. Carbon formed within the catalyst can result in its disintegration and ultimately, blocking of the tube. A molar ratio of steam to hydrocarbon of about 3:1 is typical for natural gas reforming, but in some circumstances carbon can be deposited even though this overall ratio is apparently maintained.

For instance, the presence of 'slugs' of heavier hydrocarbons in the feed with a large demand for steam can result in carbon deposition that can cause problems unless an appropriate catalyst is used. Another cause of carbon deposition is found in very highly loaded reformers, where relatively large amounts of unreacted hydrocarbon can be exposed to temperatures sufficiently high to cause thermal cracking. This can happen when insufficient reforming has taken place in the cooler, inlet part of the tube to generate sufficient hydrogen to suppress thermodynamically the cracking reactions when the gas moves into a hotter region. This situation is usually reached at the point of maximum heat transfer into the tube, about 3 m from the inlet, where the carbon formed reduces heat transfer causing a high temperature region or 'hot band' that can damage the tube metal.

There are two approaches to curing this problem, and both involve using alternative catalysts, at least in the upper, inlet, part of the tubes (conventional steam reformers are usually down flow). The first approach is to use a higher activity catalyst so that in the cooler inlet region more hydrocarbon reforming takes place, so producing more hydrogen in this part of the tube. If sufficient hydrogen is produced quickly the composition of the gas will have moved out of the thermodynamic region for carbon forming by the time it reaches the hotter parts of the tube. This approach can work but, unfortunately, catalyst activity inevitably falls with time, and once high activity is lost, carbon-producing hot bands can form. The second, and preferred, approach to countering carbon formation is to make use of special catalysts containing alkali, which promotes the reaction between carbon and steam, equation (1). The alkali has the feature of being sufficiently mobile to prevent carbon forming anywhere in the tube; more-over, it neutralizes the acidic sites on the support that encourage hydrocarbon cracking. As long as the rate of the 'carbon removal' reaction is faster than those reactions tending to form carbon, there will be no deposition of carbon. A particular advantage of the alkalized catalyst is that should carbon be formed through some malfunction, removing it with steam may be possible and often this catalyst will itself recover from plant upsets where alternative catalysts could not.

Carbon formation in naphtha steam reforming. More care is necessary in the choice of catalyst used for steam reforming hydrocarbons higher than natural gas because of their increased tendency to form carbon through cracking, and special catalyst formulations are required.[16,17] The most successful commercial catalyst for the reforming of naphtha is made by precipitation techniques rather than by impregnation of a preformed support, and it contains an alkali metal promoter that plays a complex role in preventing carbon formation. As in the hot-band problems discussed previously, the alkali neutralizes acid sites on the catalyst support that can facilitate cracking, and at the high temperatures used it catalyses the reaction of carbon with steam to give carbon monoxide and hydrogen.

3.3.4 Steam reforming catalyst poisons. Nickel steam reforming catalysts are particularly sensitive to poisoning by sulphur and arsenic, although at the temperatures used in most reformers sulphur does not completely deactivate the catalyst. Nevertheless, it is important that the sulphur compounds in the feed gas are reduced to very low levels. Natural gas contains hydrogen sulphide and low boiling sulphides or mercaptans, while naphtha contains more complex organic sulphur compounds. Typically the amount of sulphur present is in the range 50–600 ppm.

Sour gas—i.e. natural gas containing high levels of hydrogen sulphide (and simple organic sulphur compounds)—must first be subjected to an 'acid gas' removal process which removes most of the hydrogen sulphide by dissolution in a liquid. This is then followed by addition of some hydrogen and passage through a bed containing cobalt molybdate hydrodesulphurization catalyst (Co and Mo sulphides on alumina*) at $\sim$ 400°C. Organosulphur compounds are reduced to hydrogen sulphide which is removed from the process stream by adsorption/reaction on high surface area zinc oxide as illustrated in equations (8) and (9).

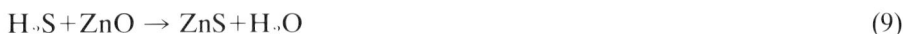

$$R_2S + 2H_2 \rightarrow H_2S + 2RH \tag{8}$$

$$H_2S + ZnO \rightarrow ZnS + H_2O \tag{9}$$

When used in a natural gas duty, cobalt molybdate hydrodesulphurization catalysts can have a life of several years, but in some situations a carbonaceous deposit accumulates during use. This results from cracking some hydrocarbon (particularly unsaturated compounds) and causes a progressive decrease in catalyst activity. However, by maintaining a minimum hydrogen partial pressure and operating below a set maximum temperature this problem can be avoided, and at the same time unsaturated hydrocarbons are hydrogenated.

3.4   Achieving composition requirement
Depending on the final use of the synthesis gas it is necessary to change the composition of the raw gas to a greater or lesser extent.

3.4.1   Gas for methanol synthesis. The synthesis gas derived from steam reforming naphtha has a carbon to hydrogen ratio close to that of methanol, and in a methanol plant based on naphtha feedstock it is not necessary to modify the composition of the synthesis gas before the synthesis stage. The situation is different when a methane-rich gas is the feedstock, since hydrogen-rich gas is produced, as illustrated in the hypothetical equation $CH_4 + H_2O \rightarrow CH_3OH + H_2$. In some plants this excess hydrogen is burnt as fuel in the reformer, but in economic terms it is better to add addition carbon dioxide to the process (obtained from flue gas or an ammonia plant) and use the excess hydrogen to make more methanol.

3.4.2   Gas for ammonia synthesis. In an ammonia plant achieving the required 3:1 hydrogen/nitrogen mixture is more complex than in a methanol plant, and requires several catalytic stages.

*In some instances nickel molybdate catalyst is used.

Secondary reforming. The nitrogen required for ammonia production is usually added to the process stream in a second reforming stage. Like 'primary reforming' discussed in the previous section, 'secondary reforming' uses a nickel catalyst most often in the form of Raschig rings that must be particularly thermally stable because of the very high temperatures employed. Air is admitted through a special burner and to prevent flame playing directly on the catalyst bed it is usual to protect it by a layer of heat-resistant alumina. The vessel itself is a well insulated adiabatic reactor that is lined with ceramic material.[18] As a portion of the process gas is combusted, the temperature, already high from the previous stage, rises to over 1100°C which drives the reforming reaction almost to completion with the exit methane level being reduced to less than 0.5%.

In spite of the very high temperatures involved there are generally few problems with secondary steam reforming catalysts either in terms of activity or life. Plant maloperation can result in catalyst breakage with resultant pressure drop problems, but most problems are associated with poor burner performance rather than with the catalyst. Most often preformed impregnated ceramic support is used, but specially formulated precipitated catalysts can also operate successfully with long lives. Nickel contents are usually rather lower than in primary reforming catalysts.

Water gas shift: carbon monoxide conversion. The water gas shift reaction, equation (3), provides an important means of converting carbon monoxide and steam to hydrogen and carbon dioxide, enabling the full potential of hydrogen production from synthesis gas to be achieved. As a consequence of the exothermic nature of the shift reaction its equilibrium constant decreases with increasing temperature, so to obtain maximum conversion of carbon monoxide into hydrogen the equilibrium should be established at the lowest possible temperature (Fig. 4.5). At a given temperature the maximum achievable conversion is increased by increasing the partial pressure of steam, but not appreciably by changing the total pressure.

The rate of the homogeneous gas phase shift reaction is very slow at convenient temperatures and in practice it is necessary to use a catalyst, but it is important that those used are selective. Catalysts with high hydrogenation activity might form methane which not only would consume valuable hydrogen and produce additional 'inerts' (see section dealing with ammonia synthesis), but the accompanying heat of reaction could take the form of dangerous thermal 'runaways'. Suitably selective catalysts based on iron and chromium oxides (the active phase being Fe_3O_4 stabilized with Cr_2O_3) have been used for more than 60 years. Until the introduction of steam reforming, synthesis gas was comparatively impure and iron/chrome catalysts gained acceptance because they are not very sensitive to poisons. However, these catalysts are not very active and must

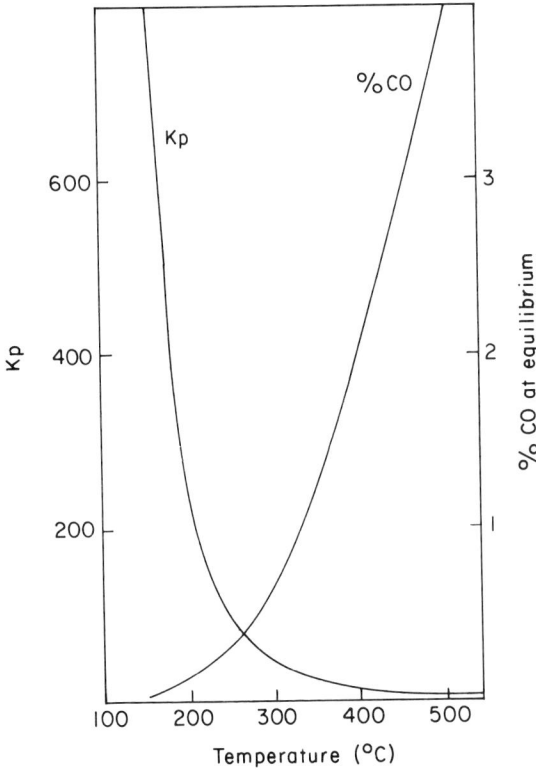

Fig. 4.5. Temperature dependence of the water gas shift reaction and the calculated equilibrium percentage of carbon monoxide in a typical gas mixture exit the low temperature shift reaction.

therefore be operated at high temperatures (e.g. $\sim 450°C$) and because of this they are known as 'high temperature shift catalysts'. Consequently, the level of carbon monoxide exit a high temperature shift reaction is around 4%.

High temperature shift catalyst is still used in modern single-stream ammonia plants, and because pure gas is used, it is also possible to use the more active copper-based catalyst capable of operating at low temperatures so ensuring almost complete conversion of carbon monoxide into hydrogen. Even with a high activity catalyst operating under ideal conditions it is not possible to achieve high conversion of carbon monoxide into hydrogen with a single fixed adiabatic bed of catalyst. This is because the minimum inlet temperature to the reactor is dictated by the dew-point under operating conditions (typically $\sim 200°C$), and because the heat evolved from the reaction increases the operating temperature in proportion to the amount of carbon monoxide converted into hydrogen. Thus, depending on plant design and operating conditions, in single stage operation the temperature rise could approach 100°C. As a result, exit carbon monoxide levels up to $\sim 1\%$ would result. Much better conversion (exit CO 0.1–0.3%) is obtained by carrying out the reaction in two stages with interbed cooling: one operating at high temperature, and the second at low temperature with a high activity catalyst.[19]

Like the high temperature shift catalyst (Fe_3O_4/Cr_2O_3) the low temperature shift catalyst is made by precipitation techniques. The very early low temperature shift catalysts contained copper and zinc oxide and some formulations, whilst having good initial activity, quickly lost activity in use due to facile sintering of the copper phase. The coalescence of the many fine copper crystals into larger ones decreased their surface area and hence activity. Addition of alumina to the formulation helped to prevent the copper sintering and significantly prolonged activity. However, considerable care is necessary in designing low temperature shift catalyst to obtain optimum performance.[20] Catalyst strength, as with other catalysts, is important. The formulation has to include sufficient alumina of an appropriate size to stabilize the copper crystallites—the 'stable activity' of a high copper content catalyst is less than that of a well formulated product containing less copper. Physical properties other than strength are also important. For instance, in terms of activity, pore volume is as important as is copper surface area, because under operating conditions the low temperature shift reaction is pore diffusion limited; the rate being proportional to the square root of the product of copper surface area and pore volume.

Catalyst poisons. Previously it was noted that high temperature iron/chrome catalysts are not particularly sensitive to poisons. However, copper catalysts operating at low temperatures are very susceptible to poisons. Sulphur- and chlorine-containing compounds have been identified as being the two most important generally encountered poisons.[21] Catalysts containing uncombined or 'free' zinc oxide enable the catalyst to trap sulphur in the top part of the charge. Because of the high affinity of the catalyst for sulphur it stays in this top part of the bed thus 'protecting', or 'guarding' the rest of the charge. As a result of this self-protecting action, under normal operating conditions, a charge of well formulated low temperature shift catalyst ought not to have its performance seriously impaired by the low levels of sulphur in synthesis gas derived from steam reforming.

In most ammonia plants the amount of chlorine (present as hydrogen chloride) in process gas is very low indeed, below the level of analytical detection. However, it is a most powerful poison for copper catalysts. It accelerates the sintering of the copper crystallites and even small amounts of adsorbed halide cause them to increase in size—perhaps by an order of magnitude. Copper surface area is then reduced and hence the catalyst's activity. The resistance of an optimized well formulated catalyst to thermal sintering helps it to resist these effects and its porous nature ensures rapid capture of hydrogen chloride in the upper part of the charge. Copper chlorides are much more 'mobile' than are the sulphides, and generally chlorine is the more serious poison. Consequently it is often advantageous to use a chlorine guard bed of alkalized alumina granules.

Economic aspects of the shift reaction. The introduction of a low temperature shift catalyst stage into ammonia plants during the 1960s significantly improved overall plant economics. This is because every mole of carbon monoxide converted produces one mole of hydrogen directly and also prevents the loss of three moles of hydrogen in the methanator (see following section). A further important saving is that purge from the synthesis loop is significantly decreased because the methane content of make-up gas is lower. The overall effect is that extra ammonia made in a 1000 tonnes day^{-1} plant resulting from the reduction of 0.1% carbon monoxide in the exit gas of the low temperature shift is the equivalent of at least 11 tonnes of ammonia per day.

Methanation: general. Carbon dioxide is removed from the process stream in a single stream ammonia plant after the low temperature shift stage. Liquid systems are used, the most common involving the carbonate/bicarbonate equilibrium in aqueous solution.[22] The resulting carbon dioxide and carbon monoxide levels in the process gas are then usually less than 0.5% each.

The purpose of the methanation stage is to remove carbon oxides from ammonia synthesis gas to low parts per million levels, because carbon monoxide and carbon dioxide are poisons for iron-based ammonia synthesis catalyst (see below). Even at these low levels of carbon oxides there is evidence suggesting long-term activity of the synthesis catalyst is improved by operation at reduced carbon oxide levels.

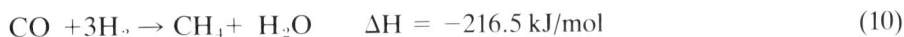

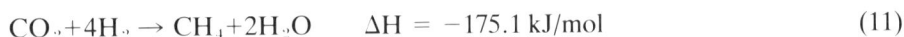

$$CO + 3H_2 \rightarrow CH_4 + H_2O \qquad \Delta H = -216.5 \text{ kJ/mol} \qquad (10)$$

$$CO_2 + 4H_2 \rightarrow CH_4 + 2H_2O \qquad \Delta H = -175.1 \text{ kJ/mol} \qquad (11)$$

Both methanation reactions are strongly exothermic and the temperature rise in a lagged adiabatic reactor can be used to estimate the inlet level of carbon oxides (74°C per 1% CO, and 60°C per 1% CO_2).

Methanation catalysts. Many transition metals are active in methanation. Early work concentrated on nickel and iron catalysts.[23] Iron catalysts do not have sufficient selectivity for methane. They form undesirable carbon and some higher hydrocarbons as a liquid product (i.e. Fischer–Tropsch products). Precious metals are also active and ruthenium catalyst has found very limited commercial application.

Industrial methanation catalysts are essentially nickel metal dispersed on a porous carrier, e.g. alumina or calcium aluminate, supplied in the form of small pellets or granules. They are generally of the precipitated type, although one supplier has an impregnated version. A good methanation catalyst is physically strong, reducible at 300°C or below and has high activity. These properties should be retained in use, and lives of up to 10 years are known.

The reduction of nickel oxide by hydrogen is almost thermoneutral at 300°C and reduction usually does not cause any problems. Once some metallic nickel has been formed by reduction with process gas, methanation will begin and produce a corresponding temperature rise. For this reason the gas used for reduction should ideally contain as little carbon oxides as possible.

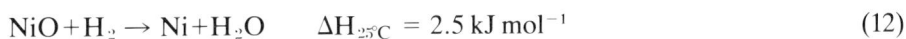

$$NiO + H_2 \rightarrow Ni + H_2O \qquad \Delta H_{25°C} = 2.5 \text{ kJ mol}^{-1} \qquad (12)$$

Poisons. With a well formulated methanation catalyst, sintering is not an important cause of activity loss at normal operating temperatures (250–350°C), even if the catalyst is occasionally overheated to ~500°C. Poisoning is the main cause of loss of activity.[21] The preceding low temperature shift catalyst is an efficient trap for poisons in the synthesis gas and the carbon dioxide removal system is the usual source of poisons causing problems with methanation catalyst. Carry-over of a small amount of liquid into the methanator is not normally serious, but plant malfunctions can sometimes result in large quantities of carbon dioxide removal liquors being put onto the catalyst. This can be serious if the liquid contains arsenic or sulphur since irreversible loss of activity results, but the effects are less important if only potassium carbonate and/or organic solvents are involved. In the former case with some mechanically strong catalysts it is possible to wash the catalyst with water and put the charge back on line.

4   Synthesis reactions

As can be seen from the previous discussion, by far the largest part of the process for the production of ammonia is the generation of ammonia synthesis gas, both in terms of capital employed and catalysts required. The synthesis reactions for ammonia and methanol are similar, in that a relatively low percentage of product exists in the gas stream leaving the converter, and substantial recycle of synthesis gas is required after removal of the product. Build-up of inert materials such as methane and argon is avoided by taking a purge from the recycle gas and recovering the fuel value.

4.1   *Ammonia*

4.1.1   Thermodynamics and kinetics. The synthesis of ammonia from nitrogen and hydrogen may be represented by the equation:

$$\tfrac{1}{2}N_2 + \tfrac{3}{2}H_2 \rightleftharpoons NH_3 \qquad \Delta H = -46.0 \text{ kJ mol}^{-1} \qquad (13)$$

Assuming the components of the mixture behave as ideal gases, the equilibrium constant may be expressed as:

$$K_p = \frac{p\,NH_3}{(p\,N_2)^{1/2}\,(p\,H_2)^{3/2}} \tag{14}$$

The variation of equilibrium constant (K_p) with temperature has been calculated from standard thermodynamic data[24] and for the temperature range 375–535°C can be expressed as in equation (15),

$$\log K_p = -5.963 + 2740/T \tag{15}$$

where T is the absolute temperature. The mole fraction of ammonia at equilibrium in a particular gas mixture at a given temperature is given by equation (16),

$$x_{NH_3}\,p = K_p\,(x_{N_2}\,p)^{1/2}\,(x_{H_2}\,p)^{3/2} \tag{16}$$

where x is the mole fraction of each component and p is the total pressure in the system. Equilibrium concentrations of ammonia derived using this equation are shown in Fig. 4.6 which assumed 10% inert material (methane and argon) in stoichiometric ammonia synthesis gas.

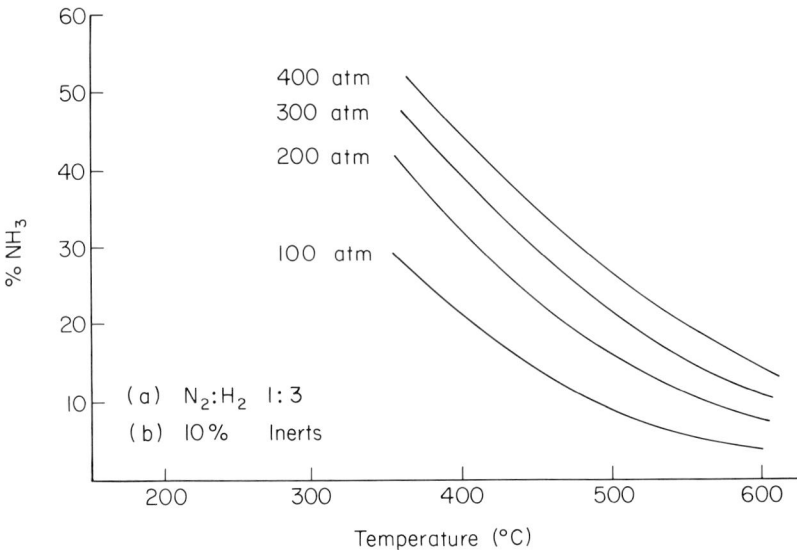

Fig. 4.6. Variation of ammonia concentration at equilibrium with temperature and pressure.

The kinetics of ammonia synthesis have been extensively studied in flow systems since about 1935, mainly by groups in the Soviet Union, Germany and

America. The work of Temkin and Pyzhev[25] in 1940 enabled a satisfactory kinetic interpretation of the synthesis reaction over doubly promoted iron to be made. For a given temperature, the synthesis rate can be expressed in general form by the equation (17),[26]

$$\frac{dp_{NH_3}}{dt} = k_1 \cdot p_{N_2} \left[\frac{p_{H_2}^3}{p_{NH_3}} \right]^n - k_2 \left[\frac{p_{NH_3}^2}{p_{H_2}} \right]^{1-n} \tag{17}$$

where k_1 and k_2 are the forward reverse rate constants. The magnitude of the constant, n, reflects how the differential heats of adsorption of nitrogen on the catalyst and the differential activation energies of adsorption and desorption of nitrogen vary with overall surface coverage of nitrogen. Reported values of n vary between 0.5 and 0.75, with the most reliable figure of 0.7 being proposed by Brill.[26]

Ammonia synthesis presents no problems arising through selectivity. The reaction is favoured by high pressure, which increases both the reaction rate and the equilibrium percentage of ammonia. The rate of reaction is increased at higher temperature, but the equilibrium percentage of ammonia is decreased. In addition, thermal sintering and hence loss of activity of the catalyst also occurs at higher temperature. In practice, current ammonia synthesis converters operate around 400–500°C and 150–350 atmospheres pressure. In the latest technology, loop pressures of 70–80 bar are indicated.[27]

The maximum theoretical level of ammonia in a gas stream containing $\sim 10\%$ inerts at 450°C and 150 atmospheres pressure is about 19%. At practicable space velocities the attained level of ammonia tends to be about 14%. The gas stream exit of the synthesis converter is cooled to $-12°C$ and about 80% of the ammonia is condensed out. A purge, to control the level of methane and argon in the stripped gas is removed and the rest is heated to synthesis temperature with fresh additional ammonia synthesis gas before recycle to the converter.

4.1.2 Catalyst formulation. The actual commercial catalyst has not changed significantly since the early days of Mittasch. He developed the potassium/aluminium promoted catalyst and the best method for making it. The process reuired fusion of magnetite, alumina and potash in the correct proportions, cooling the melt by pouring into trays and comminution of the product to the appropriate size. Modern catalysts differ only in the types of promoter added. Potash and alumina are still universally used but calcium oxide, magnesia, silica, titania, zirconia and vanadium pentoxide may be added in small quantities, prior to fusion.

During the commissioning stages of a new charge of catalyst, the magnetite component is reduced to iron metal under carefully controlled conditions to maximize the iron surface area. Reduction is generally carried out using ammonia synthesis gas at about 100 atmospheres pressure, beginning at about 350°C and gradually increasing the temperature to 450°C. Catalyst activity is impaired by contact of water with freshly reduced iron. Presumably a high partial pressure of water vapour reoxidizes the iron to magnetite leading to loss of area as the reduction proceeds to completion. As soon as some iron is reduced ammonia synthesis takes place on this new surface, leading to a considerable temperature increase due to the reaction exotherm. Thus, the gas rate needs to be carefully controlled to avoid thermal hot spots and to avoid back diffusion of wet reducing gases to the fresh iron surface. The water from the reduction step is removed from the synthesis gas by cooling and a dilute aqueous ammonia solution is obtained.

Much research effort has gone into elucidating the role of the promoters in ammonia synthesis. Alumina, typically around the 2% level, dissolves in the magnetite forming a solid solution. On reduction, the alumina comes out of solution, acting as a skeletal support for iron crystallites, thus generating significant surface area. Surface areas in the range 15–20 m^2 g^{-1} can be obtained, compared with ~1 m^2 g^{-1} for unpromoted iron. Potash added in the form of potassium carbonate in the fusion process behaves quite differently. It does not form a solid solution in the magnetite lattice, but becomes largely associated with the alumina and silica to form aluminates and aluminosilicates. Some potassium ferrites are also formed. Surface science studies[28] have shown that the ammonia synthesis reaction on metallic iron is structure sensitive and proceeds more rapidly on the Fe(111) face than the (110) with unpromoted iron. Potash interacts with acid sites on the alumina skeleton, but also activates the other iron crystallite faces to a level similar to that of the (111) face.[29] Too much potash is detrimental to catalyst performance and the optimum level is below 1%.

Iron was established as the most effective ammonia synthesis catalyst after an exhaustive search covering the periodic table. Both osmium and uranium are at least as effective as unpromoted iron with better stability, but suffered from cost and limited availability. The recognition of health problems associated with radioactivity and the highly toxic osmium gave further incentive to study the promoted iron system. Other metals such as cerium, tungsten, manganese, ruthenium and in particular molybdenum also show some activity, but were found not to be promoted by additives to the same extent as iron and were eventually discarded as candidates. However, recent work at BP[30] has shown that high activities can be obtained using ruthenium supported on carbon catalysts.

4.1.3 Catalyst deactivation. Iron-based ammonia synthesis catalysts are relatively long-lived. A typical charge of about 100 tonne catalyst in a 1000 tonne day^{-1} plant can last up to 10 years, though in practice, the catalyst may be changed more frequently to maintain high output. Loss of activity generally occurs through two mechanisms. Thermal sintering is a slow on-going process, gradually leading to loss of surface area. This is promoted by poor thermal distribution in a converter giving hot spots. The other cause of loss of activity is poisoning. Oxygenated compounds such as water, carbon monoxide and carbon dioxide are well known poisons. However, deactivation caused by exposure to these com-pounds is reversible provided the concentration is low and that the period of exposure is restricted to a few days. More 'traditional' poisons such as sulphur, chlorine, phosphorus and arsenic cause irreversible deactivation. Since their effect is cumulative, stringent purification of reaction gases is essential. For example, sulphur-free oils must be used during compression stages.

4.2   Methanol

4.2.1   Thermodynamics and kinetics. Much of the early published work on methanol synthesis deals with the reaction of carbon monoxide and hydrogen alone shown in equation (18).

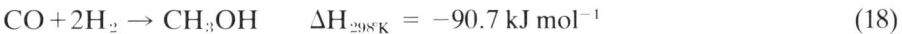

$$CO + 2H_2 \rightarrow CH_3OH \qquad \Delta H_{298K} = -90.7 \text{ kJ mol}^{-1} \qquad (18)$$

In practice, carbon dioxide is also used as a carbon source and the thermo-dynamics of methanol synthesis are complicated by water formation during synthesis from carbon dioxide as in equation (19).

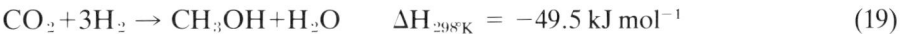

$$CO_2 + 3H_2 \rightarrow CH_3OH + H_2O \qquad \Delta H_{298K} = -49.5 \text{ kJ mol}^{-1} \qquad (19)$$

Though mechanistically, methanol has recently been shown to arise directly from carbon dioxide[31], it has been common practice to treat methanol synthesis in a gas mixture containing both carbon monoxide and dioxide as the sum of the *reverse* water gas shift reaction and the synthesis reaction between carbon monoxide and hydrogen. The equilibrium conversion of carbon monoxide to methanol is given by the expression in equation (20).

$$Ka_{(synthesis)} = \left[\frac{N_{CH_3OH} \cdot (N_T)^2}{N_{CO} (N_{H_2})^2 P^2} \right] \cdot \left[\frac{\gamma_{CH_3OH}}{\gamma_{CO} \cdot (\gamma_{H_2})^2} \right] \qquad (20)$$

N is the number of moles of each component, N_T is the total number of moles in the mixture and γ is the activity coefficient of each species. The equilibrium concentrations of the synthesis gases is given by the expression in equation (21), and the overall picture is obtained by combining these expressions.

$$Ka_{(shift)} = \left[\frac{N_{CO} \cdot N_{H_2O}}{N_{CO_2} \cdot N_{H_2}} \right] \cdot \left[\frac{\gamma_{CO} \cdot \gamma_{H_2O}}{\gamma_{CO_2} \cdot \gamma_{H_2}} \right] \qquad (21)$$

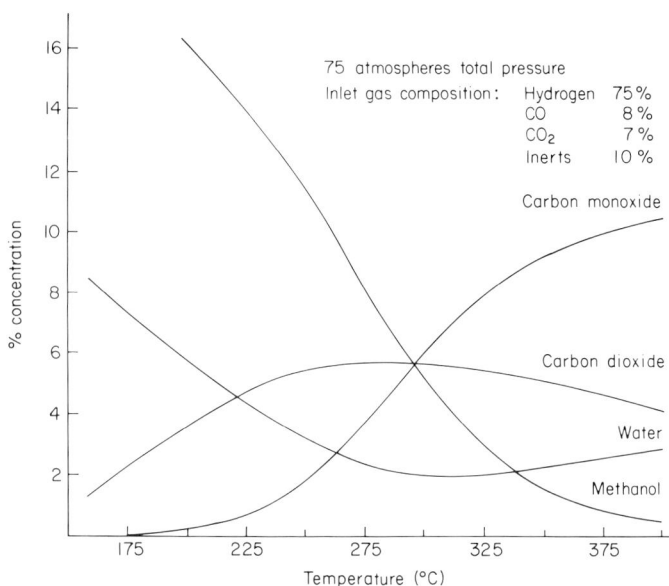

Fig. 4.7. Temperature dependence of methanol, carbon monoxide, carbon dioxide and water concentration at equilibrium.

Equilibrium concentrations for a typical gas mix are given in Fig. 4.7. Methanol formation is thus favoured at low temperatures, though clearly there is a penalty in reaction rate. Methanol formation is also favoured at high pressure. The original zinc oxide/chromia catalyst was only active at temperatures above 350°C and consequently relatively high pressures were needed. The development of catalysts active at temperatures 100°C lower enabled major savings to be made by operating a new low pressure process. A full treatment of the thermodynamics of methanol synthesis is available in the literature.[32]

As in the case of the thermodynamics, much of the kinetic data deals with synthesis gas free from carbon dioxide and water.[33] This is not relevant to current methanol technology since both water and carbon dioxide have been shown to have a major influence on the rate of reaction over $Cu/Zn/Al_2O_3$ oxide catalysts.[34] A full kinetic analysis of this reaction has still to be published.

4.2.2 Catalyst formulation. In contrast with the ammonia synthesis reaction, methanol synthesis encounters significant difficulties in selectivity. Depending upon conditions, reaction of carbon monoxide with hydrogen can give methane, polymethylene, hydrocarbons, alcohols and other oxygenated products in addition to methanol. The early plants based on zinc oxide/chromia catalysts produced liquid methanol at about 95% purity, with significant methanation also occurring. Modern plants using copper/zinc/alumina catalysts are generally about 99.5% selective to methanol, provided the iron level in the catalyst is kept to a low level. The major impurities are ethanol, higher alcohols/hydrocarbons and methyl formate, but traces of amines and unsaturated compounds can also be present.

The more modern copper/zinc/alumina catalysts are generally prepared by precipitation of mixed carbonates and ageing the gel under conditions to promote the formation of malachite containing some zinc in solid solution $(Cu, Zn)_2(OH)_2CO_3$. This material is then calcined to mixed oxides and formed into pellets using graphite lubricant. The catalyst at this stage has a surface area in the range $60-100$ m^2 g^{-1}. The catalyst is activated by controlled reduction from temperatures of 150°C to the operating temperature using 5% hydrogen in nitrogen, care being taken to avoid hot-spots in the converter which could lead to premature deactivation. When reduction is complete, as seen by no further water being evolved from the catalyst and equal inlet and exit hydrogen levels, process gas can be brought on line. After this treatment, a typical catalyst has a copper metal surface area, as measured by nitrous oxide decomposition, of about 30 m^2 g^{-1}. The structure of the catalyst, and how this relates to methanol synthesis has been studied in depth during recent years, particularly by Professor Klier's group at Bethlehem, Pennsylvania. They suggested that the active species in methanol synthesis is a solution of copper, probably Cu^I, in the zinc oxide component[35]. A review describing the state of the art in 1983 of methanol synthesis has been published.[36] However, the overall mechanism is still uncertain, as shown in recent work[37] where methanol was shown to arise directly from carbon dioxide, and the activity of the catalyst was directly proportional to the copper metal surface area.

4.2.3 Catalyst deactivation. Copper/zinc/alumina catalysts are very prone to poisoning and it is no coincidence that their development followed a change in feedstock in the industry from coal, firstly to naphtha and then to natural gas, which resulted in considerably lower sulphur levels in the process gas. The other commonly encountered poison for these catalysts is chloride and consequently the water used in the reforming stage must also be rigorously purified. The role of these poisons is complex and includes blocking of active sites, chemically-induced sintering leading to loss of metal area and expulsion of copper from the zinc oxide lattice. High temperatures are also detrimental to the catalyst. Catalyst maldistribution can cause uneven flow through a converter resulting in high temperatures and not only deactivation of the catalyst, but also higher levels of impurity in the product. Methanol catalysts tend to last for about 3–5 years on line, during which time the copper metal surface area decreases from about 30 m^2 g^{-1} to about 10 m^2 g^{-1}.

4.2.4 Recent developments. In recent years, two novel approaches to copper-catalysed methanol synthesis have been proposed. Raney copper, prepared by leaching ternary copper/aluminium/zinc alloys with strong sodium hydroxide, has shown significant activity for methanol synthesis.[38] More surprisingly, the inter-metallic compounds $ThCu_6$, $ThCu_{3+6}$, $ThCu_2$ and Th_2Cu all react directly with synthesis gas giving a form of copper on thoria which is extremely active for methanol synthesis.[39] Carbon dioxide is not essential for this catalyst and thus there is potential for a low pressure process generating anhydrous methanol directly. Since the discovery of the copper/thorium catalysts, similar intermetallic compounds based on non-radioactive rare earth metals have been shown to yield surprisingly active catalysts for the synthesis of methanol, operating at tempera-tures as low as 70°C in gas mixtures substantially free from carbon dioxide.[40]

Catalysis using precious metals has generally suffered from poor selectivity[41] to methanol. The discovery that palladium,[42] under certain conditions, could be selective to methanol stimulated activity in several research establishments.[43] However, problems associated with the low rate of reaction, the high cost of palladium and the relatively high reaction temperatures leading to some methanation, have led to reduced interest in this topic, and the copper/zinc/alumina catalyst is still the all-round best available.

5   Conclusions

The industrial processes for the production of ammonia and methanol have changed significantly since their introduction earlier this century. Many of these changes have been introduced gradually, resulting from the availability of improved materials (e.g. high performance alloys) and mechanical engineering advances such as the adoption of centrifugal compressors. This improved tech-nology has proved to be sufficiently flexible to adjust to new feedstocks, coping with resulting demands smoothly and positively. A result of such changes in the processes has been increasing efficiency and the modern plant is highly optimized in terms of energy usage. These technical improvements centre around the operation of several catalytic stages where development of the catalysts involved has been of fundamental importance. For instance, the introduction of copper-based low temperature shift and of methanol synthesis catalysts has had major impact on the economics of ammonia and methanol production. Future develop-ments of the conventional process are likely to be relatively modest, concentrating mainly on capital saving measures and further improvements in energy efficiency. A trend to lower operating temperatures and pressures is already apparent, but this is bound by chemical engineering constraints. However, such further improvements will be dependent upon continual development of the catalysts themselves.

6   References

1   Ramsay, W. & Young, S. *J. Chem. Soc.* 1884, **45**, 88.
2   Perman, W. & Atkinson, G. *Proc. Roy. Soc., London*, 1904, **74**, 110 and 1905, **76**, 167.
3   Nernst, W., Jost, W. & Jellinek, G. *Z. Elektrochem.* 1907, **13**, 152.
4   'Haber Memorial Lecture' by J.E. Coates. In *Memorial Lectures Delivered Before The Chemical Society*, The Chemical Society (London), 1951, p. 127.

5   Mittasch, A. *Advances in Catalysis*, 1950, **2**, 81.
6   *Im Banne der Chemie, Carl Bosch, Leben und Werk*, by K. Holdermann, Econ-Verlag, 1954.
7   *Nobel Lectures (Chemistry)*, Vols 1901–21 and 1922–42, Elsevier, Amsterdam, 1966.
8   Sabatier, P. & Senderens, R. *Comptes. Rend.* 1901, **132**, 1257.
9   Dry, M.E. *Applied Industrial Catalysts* 1983, **2**, 167.
10 Mittasch, A., Pier, M. & Winkler, K. Canadian Patent 251,484, (1925, to BASF).
11 Kotowski, W. *Chem. Tech.* 1963, **15** (4), 204.
12 Bridger, G.W. *Catalysis* SPR, Chemical Society (London), 1980, **3**, 39.
13 Rostrup-Nielsen, J.R. *Catalysis Science and Technology* 1984, **5**, 1.
14 Twigg, M.V. In *Catalysis and Chemical Processes*, R. Pearce & W.R. Patterson (Eds), p. 11. Leonard Hill, 1981.
15 Dent, F.J., Moignard, L.A., Eastwood, A.H., Blackburn, W.H. & Hebden, H. *Trans. Inst. Gas. Eng.* 1945–6, 602.
16 Andrew, S.P.S. In *Ammonia* Part I, A.V. Slack & G.R. James (Eds), p. 175. Marcel Dekker, New York, 1973.
17 Davies, P. & Devereux, J.M. UK 953,877 (1964, to ICI).
18 Nobles, E.J. In *Ammonia* Part 1, A.V. Slack & G.R. James (Eds), p. 275. Marcel Dekker, New York, 1973.
19 Rase, H.F. *Chemical Reactor Design for Process Plants*, Vol. II, p. 44. J. Wiley and Son, New York, 1977.
20 Lloyd, L. & Twigg, M.V. *Nitrogen* 1979 (118), 30.
21 Denny, P.J. & Twigg, M.V. In *Catalyst Deactivation*, B. Delmon & G.F. Froment (Eds), p. 577. Elsevier, Amsterdam, 1980.
22 Strelzoff, S. *Technology and Manufacture of Ammonia*, pp. 193–250. J. Wiley and Sons, New York, 1981.
23 Hilditch, T.P. *Catalytic Processes in Applied Chemistry*. Chapman and Hall, London, 1929.
24 Congdon, W.J. See Kirk-Othmer, *Encyclopedia of Chemical Technology* (2nd edn), p. 297, Ref. 12.
25 Temkin, M.I. & Pyzhev, V.M. *Acta Physiochim, URSS* 1940, **12**, 327.
26 Brill, R. *J. Chem. Phys.* 1951, **19**, 1047.
27 Livingstone, J.G. & Pinto, A. *A.I.Ch.E. Ammonia Safety Symposium*, Paper 123 F, Los Angeles, 1982.
28 Grunze, M. In *The Chemical Physics of Solid Surfaces and Heterogeneous Catalysis*, eds. D.A. King & D.P. Woodruff (Eds), Vol. 4, p. 143. Elsevier, Amsterdam, 1982.
29 Ertl, G., Lee, S.B. & Weiss, M. *Surface Science* 1982, **114**, 527.
30 McCarrol, J.J. & Tennison, S.R. UK Patent 2,109,731 (1981, to BP).
31 Kagan, Yu. B., Rozovskii, A. Ya, Liberov, L.G., Slivinskii, E.V., Lin. G.I., Loktev, S.M. & Bashkirov, A.N. *Doklady Acad. Nauk. SSSR*, 1975, **224** (5), 1081. [English Translation 1976, 598.]
32 Wade, L.E. Gengelbach, R.B., Trumbley, J.L. & Hallbauer, W.L. Kirk-Othmer *Encyclopedia of Chemical Technology* (3rd edn), pp. 403–407. 1981.
33 Kung, H.H. *Catal. Rev. Sci. Eng.* 1980, **22** (2), 235.
34 Herman, R.G., Klier, K., Simmons, G.W., Finn, B.P., & Bulko, J.B. *J. Catal.* 1979, **56**, 407.
35 Mehta, S., Simmons, G.W., Klier, K. & Herman, R.G. *J. Catal.*, 1979, **57**, 339.
36 Klier, K. *Advances in Catalysis* 1982, **31**, 243.
37 Chinchen, G.C., Denny, P.J., Parker, D.G., Short, G.D., Spencer, M.S., Waugh, K.C. & Whan, D.A. *ACS Symposium on methanol and synthetic fuels*, Philadelphia, August 1984.
38 Marsden, W.L., Wainwright, M.S. & Friedrich, J.E. *Ind. Eng. Chem.* 1980, **19**, 551.
39 Baglin, E.G., Atkinson, G.B. & Nicks, L.J. *Ind. Eng. Chem.* 1981, **20**, 88.
40 Short, G.D. & Jennings, J.R. Euro. Patent 117, 944 (1984, to ICI).
41 Ichikawa, M. *Bull. Chem. Soc. Jap.* 1978, **51** (8), 2268.
42 Poutsma, M.L., Rabo, J.A. & Risch, A.P. US Patent 4,119,656 (1978, to Union Carbide).
43 See for example, Poels, E.K., Mangnus, P.J., Welzen, J.V. & Ponec, V. *Proc. 8th Int. Cong. Catal.*, 1984, Vol. 2, 59 and references therein.

5 Industrial applications of homogeneous catalysts

Robin Whyman

1   Introduction

Other chapters in this book focus on various aspects of the uses of traditional heterogeneous catalysts in the chemical industry. Such materials comprise the major class of catalysts to find application in chemical technology. Reactants are introduced as gases or liquids but the chemical reactions take place on surfaces, i.e. there is a phase difference between the catalyst, reactants and products. An important and, in many ways, complementary area is that of homogeneous catalysis in which the catalysts—frequently metallo-organic molecules or clusters of molecules—are present in the same phase (usually liquid) as the reactants and products. Such catalysts have significant industrial applications, most of which are

128

described in this chapter, and are almost certain to find increasing use in the development of future high technology catalysts, for the reasons outlined below.

2   Homogeneous and heterogeneous catalysis

In approaching a discussion of industrial applications of homogeneous catalysis it is perhaps appropriate first to compare and contrast in general terms the properties associated with both homogeneous and heterogeneous catalysts (see Table 5.1). The thermodynamic principles of activation are of course the same in both cases but the physical and geometric factors are rather different. As amply illustrated in other chapters, heterogeneous catalysts often comprise metals or metal oxides which may be dispersed on inorganic oxide supports in order to increase the effective surface area per unit weight of metal component. Heterogeneously-catalysed reactions are usually operated at relatively high temperatures (250–900°C) by passing the reactants in the vapour phase over solid catalysts. In contrast, homogeneously-catalysed reactions are carried out in the liquid phase at considerably lower temperatures (< 250°C) in the presence of soluble metal complexes. Such reactions are frequently operated under pressure in order to increase the effective concentration of gaseous reactant(s) in solution and hence (usually) the reaction rate.

Table 5.1. Homogeneous vs. heterogeneous catalysis

	Homogeneous	Heterogeneous
Form	Soluble metal complexes, usually mononuclear	Metals, usually supported, or metal oxides
Phase	Liquid	Gas/solid
Temperature	Low (< 250°C)	High (250–900°C)
Activity	Low	High
Selectivity	High	Frequently low
Mechanisms	Reasonably well understood	Poorly understood

An inverse relationship between catalytic activity and selectivity is often apparent in both forms of catalysis. Thus, the major advantage of homogeneous catalysis is associated with the high selectivities which may be achieved, presumably because the active catalysts are restricted essentially to one type of coordination site. However, this selectivity advantage is only usually achieved at the expense of reaction rate. A notable exception to this generalization derives from the elegant work of Wilke *et al.*[1] where both high catalytic activities *and* product selectivities were observed in the dimerization of propylene catalysed by π-allyl nickel halides in combination with Lewis acids such as $EtAlCl_2$. This reaction was discovered in 1963 and very high catalytic activities of *c.* 3–6 kg product g^{-1} Ni hour^{-1} at $-20°C$ were measured. Moreover, the dimerization could be controlled within wide limits by the addition of phosphines having different electronic and, more particularly, steric properties, so that dimers of various structures could be prepared at will. By careful attention to reaction conditions, including factors such as rates of stirring, cooling and gas feed, it has recently been found that the activity is in fact about a factor of 10^3 higher, e.g. 6–7000 kg product g^{-1} Ni hour^{-1} corresponding to catalytic activities, at room temperature, such as those displayed by enzymes. Developments such as these lead to the belief that homogeneous systems hold such exciting prospects for the production of the high technology catalysts of the future.

Mechanistically, homogeneously catalysed reactions are better understood than their heterogeneous counterparts. This is mainly because they are kinetically simpler, as a result of their higher selectivities, but also because they are more amenable to studies by physical techniques, e.g. vibrational and NMR spectroscopies, under operating conditions of pressure and temperature. With heterogeneous catalysts, in contrast, there are obvious difficulties associated with the assessment of intimate mechanisms of adsorption and reaction on a surface especially under catalytic reaction conditions.

As will be clear from other chapters of this book, heterogeneous catalysts are the materials of choice for industrial processes, principally on the grounds of higher activity, but also in terms of chemical engineering requirements such as the ease of separation of products from catalyst and reactants, and robustness, i.e. the ability of the catalyst to withstand vigorous working conditions. Thus, there can be no possibility of the catalyst plating out from solution onto the reactor walls, due to instability caused by excessive temperatures or deficiencies of gases in solution, as can be a problem with homogeneous catalysts. The latter do, however, offer several potential advantages in addition to the high selectivities discussed previously. First, the catalyst is uniform, usually having only one type of reaction site which is not subject to physical surface effects and may therefore be more reproducible. Secondly, the catalyst may be readily modified in a defined manner by variation of parameters such as (i) metal (type, concentration and oxidation state), (ii) added ligands (type, ligand:metal ratio, steric and electronic effects), (iii) added anions and cations, and (iv) solvent. Thirdly, the catalyst is potentially cheaper since, in theory, all metal atoms can be used rather than an unknown fraction in heterogeneous catalysis—an important factor if precious metals are involved on a large scale. Finally, the study of kinetics and mechanism is made easier and it is sometimes possible to isolate key intermediates.

The subject of homogeneous catalysis has been discussed in many books and review articles[2] but two books which provide very appropriate background reading to this chapter are those by Parshall[3] and Masters[4]. They provide balanced descriptions of homogeneous catalytic reactions that are useful, either in the organic synthesis laboratory or in industry. All major industrial processes catalysed by soluble transition metal complexes are described. Industrial applications have also been the subject of separate reviews.[5,6] A treatise by Collman & Hegedus[7] on the principles and applications of organotransition metal chemistry contains much useful background mechanistic chemistry. Finally, a book by Pearce & Patterson[8] on applications of catalysis (both homogeneous and heterogeneous) in chemical processing provides much additional information and detail.

Clearly, the areas of homogeneous and heterogeneous catalysis are closely interlinked and in some instances the initial development of homogeneously-catalysed process has later been superseded by a heterogeneous counterpart, as in the oxidation of ethylene to vinyl acetate. In contrast, the hydroformylation reaction was originally believed to be a heterogeneously-catalysed process! Also, the majority of commercial processes comprise several stages which are operated in an integrated manner. Some steps may be homogeneously-catalysed and others may be heterogeneous. As far as catalysis by transition metals is concerned the totally homogeneously-catalysed processes currently of major significance are: the hydroformylation of olefins to aldehydes and alcohols, the oxidation of ethylene to acetaldehyde, the hydrocyanation of butadiene to adiponitrile, the carbonylation of methanol to acetic acid and, very recently, the carbonylation of methyl acetate to acetic anhydride. Several stage processes such as the Shell higher olefins process (SHOP) are believed to involve at least one homogeneously-catalysed step. All these processes relate to relatively large product tonnages and the heavy or commodity chemicals business. In contrast, an elegant demonstration of selectivity in homogeneous catalysis is provided by the application of asymmetric hydrogenation in the synthesis of L-dopa, commercialized by Monsanto. Such applications will be discussed in this chapter, together with some emphasis on the more recent developments not covered in earlier books and review articles. The first, and technically most important homogeneously-catalysed process, namely hydroformylation, will also be discussed in some detail in order to illustrate the different stages of development which led from its initial operation as a very high pressure and high temperature process to one which may now be carried out under ambient pressures and very modest temperatures. Possible future developments in areas such as C_1 chemistry will be summarized to provide a perspective for the future direction of homogeneous catalysis.

Note: In cobalt-catalysed air oxidation processes, such as the oxidation of *p*-xylene to terephthalic acid and of cyclohexane to cyclohexanone and cyclohexanol, the metal ion is believed to function simply as an initiator of the reaction rather than playing an active role in the catalysis. Such processes are therefore excluded from consideration in this chapter.

3   Hydroformylation

Hydroformylation (or the OXO reaction)[9-11] is the name used to describe reaction (1), the addition of carbon monoxide and hydrogen to the C=C double bond of

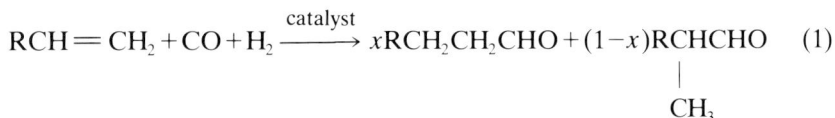

$$RCH = CH_2 + CO + H_2 \xrightarrow{\text{catalyst}} xRCH_2CH_2CHO + (1-x)RCHCHO \quad (1)$$
$$| $$
$$CH_3$$

olefins, or other unsaturated compounds, with the formation of aldehydes containing one more carbon atom than the olefin. The reaction was discovered in 1938 by Roelen at Ruhrchemie while investigating the recycling of olefins in the heterogeneously-catalysed Fischer-Tropsch synthesis. Since that time the reaction has been developed to form the basis of an industry which, worldwide, accounts for 4–5 million tonnes per year of products. The aldehydes which are initially produced in the reaction are not normally used as such, but are reduced to alcohols over heterogeneous catalysts for solvent, PVC plasticizer and detergent applications (see Table 5.2). In the case of n-butyraldehyde an important application is its conversion into a C_8 alcohol, by a base-catalysed aldol condensation, followed by dehydration and hydrogenation (reaction (2)).

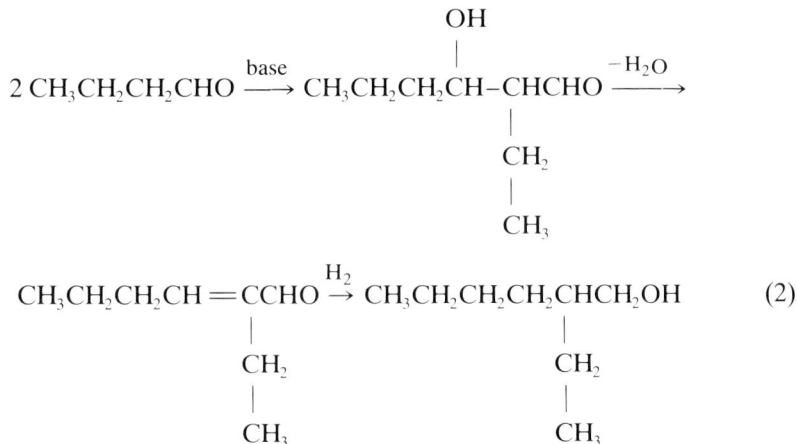

$$\begin{array}{c} & & OH \\ & & | \\ 2\ CH_3CH_2CH_2CHO \xrightarrow{\text{base}} CH_3CH_2CH_2CH-CHCHO \xrightarrow{-H_2O} \\ & & | \\ & & CH_2 \\ & & | \\ & & CH_3 \end{array}$$

$$CH_3CH_2CH_2CH = CCHO \xrightarrow{H_2} CH_3CH_2CH_2CH_2CHCH_2OH \quad (2)$$
$$\qquad\qquad | \qquad\qquad\qquad\qquad\qquad\qquad |$$
$$\qquad\qquad CH_2 \qquad\qquad\qquad\qquad\qquad\quad CH_2$$
$$\qquad\qquad | \qquad\qquad\qquad\qquad\qquad\qquad |$$
$$\qquad\qquad CH_3 \qquad\qquad\qquad\qquad\qquad\quad CH_3$$

The product molecule, 2-ethylhexanol, has desirable properties as a plasticizer alcohol. Commercially, the conversion of propylene to butyraldehydes is by far the most important application of hydroformylation technology. However, although the bulk of production via hydroformylation involves relatively simple olefins the reaction can also be applied to a great variety of substituted olefins, both simple and complex, including unsaturated oils, fats and polymers, as well as unsaturated cyclic compounds such as terpenes, pyrans, carbohydrates and

Table 5.2. Hydroformylation: most common raw materials and end-uses of products

Raw material	Hydroformylation product	Alcohol	End-use
Propylene	*iso*-butyraldehyde +	*sec*-butanol	Solvent
	n-butyraldehyde	*n*-butanol	Lacquer industry
		2-ethylhexanol	Plasticizer alcohols
n-Heptene	octanaldehyde	octanol	for PVC
Nonene (tripropylene, C₉)	C₁₀ aldehydes	'Isodecanol'	
Dodecene (tetrapropylene C₁₂)	C₁₃ aldehydes	'Tridecanol'	Synthetic detergents

steroids. In the future the reaction may well find considerable application in the fine chemicals industry, particularly where precious metal catalysts are involved.

The basic reaction is highly exothermic, with a heat of reaction for propylene of 125 kJ mol^{-1} and 115–145 kJ mol^{-1} for other alkenes, depending upon olefin structure and molecular weight, and is thermodynamically favourable at ambient pressures and low temperatures. However, it is only recently, almost 40 years after the discovery of the reaction, that these mild conditions have been realized in practice. The reaction proceeds only in the presence of metal carbonyl catalysts and is the most important industrial synthesis to use these. Although the initial catalysts used were heterogeneous, e.g. $Co(30\%)/SiO_2(66\%)/ThO_2(2\%)$ $MgO(2\%)$, the reaction is homogeneously catalysed. Indeed it was the first homogeneously-catalysed process to find industrial use, although this was neither realized nor established until several years after its discovery. There are essentially three types of olefin hydroformylation processes which have been commercialized: (i) the cobalt process, (ii) the phosphine-modified cobalt process, (iii) the phosphine-modified rhodium process, and these will be discussed in turn. The discovery and development of these processes have interesting parallels with progress in organometallic chemistry.

3.1   The cobalt process

As indicated previously the hydroformylation reaction was initially found to be catalysed by cobalt. The reaction conditions and key features of the original process are summarized in Table 5.3, from which several important points emerge. First, many forms of cobalt may be used, e.g. the carbonyl, the metal, the hydroxide, oxide, carbonate, sulphate, fatty acid salt, and Raney cobalt, all of which are believed to be converted into common cobalt carbonyls and hydridocarbonyls under reaction conditions. Secondly, the reaction conditions are fairly severe, particularly in respect of the high pressure requirement which in turn

Table 5.3. Hydroformylation reaction conditions

Classical cobalt-catalysed process

Catalyst precursor	$Co_2(CO)_8$ or Co salts
Pressure	200–300 atm
CO/H_2 ratio	1:1
Temperature	110–160°C
Catalyst concentration (% metal/olefin)	0.1–1%
n/iso ratio	4:1
Olefin hydrogenation	< 2%
High boiling products	5%

has, until recently, considerably inhibited accurate kinetic and mechanistic studies. As will be apparent from equation (1) an isomeric mixture of straight and branched chain aldehydes is produced in the hydroformylation reaction (except in the case of R = H). The composition of this isomeric mixture, which is commonly defined as the *n/iso* ratio, is one of the most important parameters of the hydroformylation process. In general, maximum selectivities to straight chain products are preferred. This is particularly important if one of the components is of higher value, as is the case with the hydroformylation of propylene to *n*- and *iso*-butyraldehyde. It is not as important if the mixtures are converted together and are equivalent in the final products, e.g. as in the conversion of diiso-butylene or tripropylene to plasticizer alcohols. A final feature to emerge from Table 5.3 concerns undesirable competing side reactions such as the hydrogenation of the starting olefin to paraffin hydrocarbon and the formation of high boiling products (termed 'heavy ends') from condensation reactions of the product aldehydes. A factor which is related to the former concerns the hydrogenation of the aldehydes to alcohols. This is not important if the entire product is subsequently hydrogenated to alcohols but, if aldehydes are the main requirement, as with *n*-butyraldehyde for subsequent conversion into 2-ethylhexanol, then low alcohol formation is required in the primary hydroformylation step. With the conventional cobalt-catalysed process, none of these side reactions pose any major difficulties.

Whereas the *n/iso* product ratio appears relatively independent of catalyst concentration and solvent it does exhibit some dependence on temperature and partial pressures of carbon monoxide, the use of lower temperatures and higher P_{CO} favouring the formation of the straight chain isomer, although with a concomitant decrease in overall conversion. These effects are probably caused by shifts in the equilibria between straight and branched chain alkyl and acyl cobalt carbonyls. By making use of these variables it was possible to exercise some control over the *n/iso* ratio, within certain limits, and to operate the cobalt-catalysed hydroformylation process successfully for *c.* 20 years without the necessity for technical innovation and improvement. One significant problem however remained, and this relates to the instability of cobalt carbonyls under certain reaction conditions (particularly high temperatures) and their tendency to deposit from solution onto the walls of the reactors. In addition to considerable loss of catalytic activity this also resulted in significant problems with catalyst recovery. Overall, therefore, a number of incentives to improve the classical cobalt-catalysed process became apparent: (i) to operate at lower pressures, (ii) to improve the *n/iso* ratio further, (iii) to enhance the stability of the catalyst under operating conditions, (iv) to improve catalyst recovery, and (v) to minimize competing side reactions.

As a first step in this direction a very obvious possibility was to investigate the catalytic activity of metal carbonyls other than cobalt. The results of such a study are summarized in Table 5.4 from which it is clear that cobalt and rhodium are the metals of choice for hydroformylation. In spite of the considerably enhanced catalytic activity displayed by rhodium it has two major disadvantages: (i) cost and (ii) it gives a *n/iso* aldehyde product ratio of only 1:1, compared to 4:1 in the case of cobalt. Other metal carbonyls are considerably less active and it is at first sight surprising that the traditional Fischer–Tropsch catalysts, iron and nickel, are so much less active than cobalt. Thus, no clear commercially attractive alternative to cobalt emerged from this study.

Table 5.4. Relative activities (in parentheses) of metal carbonyls or derivatives as hydroformylation catalysts

Mn	Fe	Co	Ni
(10^{-4})	(10^{-6})	(1)	$(< 10^{-6})$
	Ru	Rh	Pd
	(10^{-2})	(10^{3})	
Re	Os	Ir	Pt
		(10^{-2})	

In parallel with the development of organometallic chemistry during the late 1950s and early 1960s it became apparent that replacement of the carbonyl groups of metal carbonyls by ligands, particularly organophosphines and organoarsines, led to the formation of complexes of higher thermal stability. This led, in turn, to a surge of interest in the application of the chemistry of transition metal organo-metallics and hydrides to the homogeneous catalysis of the hydroformylation reaction. It is these coordination complexes, which consist of organic ligands bonded to the previously studied metal carbonyls, which form the basis of the most recent developments in hydroformylation chemistry and which have led to the appearance of two new hydroformylation processes, namely the Shell phosphine-modified cobalt process[12, 13] and the Union Carbide/Johnson Matthey/ Davy McKee phosphine-modified rhodium process.[14]

3.2   Phosphine-modified cobalt process

The key features of this process are summarized in Table 5.5 and a comparison between this and Table 5.3 clearly demonstrates both advantages and disadvantages of the phosphine-modified catalyst. Thus, the addition of the phosphine has led to the generation of a catalyst with increased thermal stability which, in turn, facilitates operation at lower pressures and higher temperatures. In addition the n/iso ratio is significantly improved by the incorporation of the phosphine ligand. However, these advantages are offset by a considerable reduction in reaction rate (at 180°C the rate is only 20% of that of the conventional cobalt-catalysed process at 145°C) and the fact that the hydrogenation capacity of the system has been enhanced to such an extent that paraffin formation becomes a serious side reaction. If alcohols rather than aldehydes are the required products then this can be a very useful process. Shell have commercialized the process to do just this—the conversion of C_8–C_{15} terminal olefins into linear alcohols. The relatively high thermal stability of the catalyst considerably facilitates the separation of products from the catalyst by distillation. A modification of this catalyst forms the basis of Shell's Aldox process for the single-step conversion of propylene into 2-ethylhexanol. Co-catalysts, such as compounds of Zn, Sn, Ti, Al, Cu or potassium hydroxide, are added to the hydroformylation catalyst and enable propylene hydroformylation, aldol condensation and hydrogenation to take place simultaneously.

Table 5.5. Phosphine-modified cobalt process

Catalyst precursor	$Co_2(CO)_8 + n\text{-Bu}_3P$
Phosphine/Co ratio	2:1
Pressure	50–100 atm
Temperature	160–200°C
Catalyst concentration (% metal/olefin)	0.6%
n/iso ratio	7:1
Olefin hydrogenation	10%
High boiling products	1%

3.3   Phosphine-modified rhodium process

The essential features of this process are summarized in Table 5.6. A comparison between this table and the data in Tables 5.3 and 5.5 indicates considerable further progress in the trend towards milder reaction conditions and increased

Table 5.6. Phosphine-modified rhodium process

Catalyst precursor	$HRh(CO)(PPh_3)_3 + PPh_3$
Phosphine/Rh ratio	50–100:1
Pressure	1–25 atm
Temperature	60–120°C
Catalyst concentration (% metal/olefin)	0.01–0.1
n/iso ratio	8:1 → 16:1
Olefin hydrogenation	5–10%
High boiling products	Low

linearity of products. With respect to the first point the reaction proceeds at 1 atm total pressure but the rates required for commercial operation are only attained at somewhat higher pressures. The *n/iso* ratio is dependent upon the amount of added phosphine but can be as high as 16:1, corresponding to a 94% selectivity to linear products. As a consequence of the mild reaction conditions only small amounts of high boiling products are obtained, although olefin hydrogenation is a significant side reaction. The catalyst concentration is low compared to both cobalt processes but against this has to be set the cost, availability and efficient recovery of rhodium (however, the catalyst is said to make only a minor contribution to production costs,[15] i.e. less than 0.5c per lb of butyraldehyde produced). This process, although excellent for the hydroformylation of lower olefins, e.g. propylene to butyraldehyde, is of limited use for higher olefins because of thermal instability of the catalyst (the high temperatures necessary for distillation of the product from the catalyst result in its decomposition). Thus, at its present stage of development, the phosphine-modified rhodium catalyst is less versatile than either of the cobalt-based processes. However, the mild reaction conditions under which this catalyst operates makes the rhodium–phosphine system potentially very attractive for the synthesis, on a small scale, of high added value chemicals.

3.4   *Factors governing the choice of hydroformylation process*

It will be clear from the previous descriptions that in many respects the three commercial processes are complementary rather then competitive, and the principal factor governing the choice of a particular process concerns the required versatility in terms of substrate carbon chain length. Thus, if *n*-butyraldehyde is the required primary product, as an intermediate in the manufacture of 2-ethylhexanol, then the phosphine-modified rhodium system is the clear process of choice. However, this process is not very versatile and if the requirement is for hydroformylation of a range of higher carbon number olefins to mixed aldehydes then the original non-liganded cobalt process is still satisfactory. If mixtures of alcohols are the desired products then the phosphine-modified cobalt process is appropriate, providing that some loss of olefin by hydrogenation to the paraffin hydrocarbon can be tolerated. Thus, the choice of a particular process is not always clear-cut and, in spite of the advantages of the phosphine-modified rhodium process in terms of mild operating conditions and high selectivities to linear aldehydes, it is worth noting that approximately 80% of the world capacity of hydroformylation plants still uses the original non-liganded cobalt process. Clearly, however, this figure will decrease as new plants come on stream and the future trend will lie in the direction of milder operating reaction conditions.

3.5   Kinetics and mechanism of hydroformylation

Two principal approaches have been taken towards understanding the hydro-formylation reaction. First, the development of rate equations based on experimental data which describe the effects of changes in variables such as metal concentration, pressure and temperature on reaction rate and product distributions. As indicated previously, this has been a difficult approach because of the severe conditions of pressure and temperature used in the initial work which led to problems of reproducibility of the data. A second approach has been to study the reactions of metal carbonyls with olefins, primarily at low pressures, in attempts to obtain evidence for plausible reaction intermediates. It is only relatively recently that these two approaches have converged, with the development of spectroscopic cells to observe the reacting species directly under process conditions. A unified overall picture of the reaction has thus emerged and the reaction is reasonably well understood. However, even now—over 45 years after the discovery of hydroformylation—some questions remain unanswered.

Although detail differences between the cobalt, cobalt–phosphine and rhodium–phosphine processes are apparent, the accumulation of extensive experimental data has led to a generally accepted reaction mechanism which, for

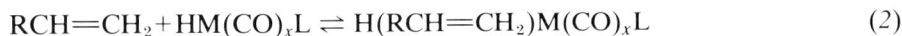

$$M_2(CO)_6L_2 + H_2 \rightleftharpoons 2HM(CO)_3L \rightleftharpoons 2HM(CO)_2L + 2CO \tag{1}$$

$$RCH=CH_2 + HM(CO)_xL \rightleftharpoons H(RCH=CH_2)M(CO)_xL \tag{2}$$

$$H(RCH=CH_2)M(CO)_xL \rightleftharpoons RCH_2CH_2M(CO)_xL \tag{3}$$

$$RCH_2CH_2M(CO)_xL + CO \rightleftharpoons RCH_2CH_2COM(CO)_xL \tag{4}$$

$$RCH_2CH_2COM(CO)_xL + H_2 \rightarrow RCH_2CH_2CHO + HM(CO)_xL \tag{5a}$$

$$RCH_2CH_2COM(CO)_xL + HM(CO)_3L \rightarrow RCH_2CH_2CHO + M_2(CO)_{3+x}L_2 \tag{5b}$$

Fig. 5.1. Hydroformylation reaction mechanism.   M = Co.   L = CO or PR$_3$.   x = 2 or 3.

the first two process catalyst precursors, is indicated in Fig. 5.1. The different steps contributing to the catalytic cycle contain many of the classical concepts developed during the last 20 years of organometallic chemistry. For example, the activation of the catalyst, step *1*, comprises the formation of an 18-electron metal carbonyl hydride species which, through dissociation of a molecule of carbon monoxide, is in equilibrium with a coordinatively unsaturated species $HCo(CO)_2L$. This equilibrium mixture is believed to form the active catalyst. Step *2* of the reaction scheme involves coordination of the reacting substrate (the olefin) and step *3* a formal 'insertion' of the olefin into the Co—H bond to form an alkyl complex. The next step in the reaction sequence involves insertion (more accurately described in mechanistic terms as an alkyl group migration reaction) of molecular carbon monoxide into the Co—C bond to give the metal acyl species, followed, step *5*, by its hydrogenolysis to product aldehyde together with regeneration of the starting catalyst and/or its precursor. It should be noted that steps *1–4* are all reversible, during which olefin, alkyl and acyl isomerization can all occur, and that there is a parallel reaction sequence for the formation of the branched chain products.

The sequence of steps *1, 2* and *3* combined, *4* and *5(a)* has been demonstrated by IR spectroscopic measurements under pressure for a closely related Ir/phosphine hydroformylation catalyst system.[16] Although the mechanism depicted in Fig. 5.1 finds general acceptance there is no direct spectroscopic evidence for the occurrence of step *2* under actual reaction conditions and the mechanism by which the acyl is converted into aldehyde remains controversial. The two possibilities, hydrogenolysis by molecular hydrogen *(5a)* or bimolecular elimination *(5b)* may both occur at comparable rates with different metal carbonyl/phosphine catalyst combinations.

In the case of the phosphine-modified rhodium process a similar overall scheme is believed to be operative although the early stages involve an associative reaction (3) rather than a dissociative process, viz:

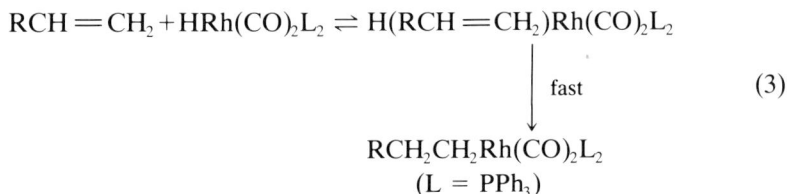

$$RCH\!=\!CH_2 + HRh(CO)_2L_2 \rightleftharpoons H(RCH\!=\!CH_2)Rh(CO)_2L_2$$

$$\Big\downarrow \text{ fast} \qquad\qquad (3)$$

$$RCH_2CH_2Rh(CO)_2L_2$$
$$(L = PPh_3)$$

An associative process would also provide more steric hindrance and favour the formation of linear products.

3.6   *Current trends and future opportunities in hydroformylation*
In the past 45 years the hydroformylation reaction has evolved from one in which

relatively severe reaction conditions were necessary, thus effectively limiting the reaction to 'industrial-only' use, to one which can be operated at ambient pressures and moderate temperatures, thus bringing the reaction within the capacity of the synthetic organic chemist. It seems likely that hydroformylation should find significant applications in the fine chemicals industry and this is an area where the next new developments are perhaps most likely to appear. An aspect of particular importance is asymmetric hydroformylation, the potential of which has yet to be realized, possibly as a consequence of the many reversible steps outlined in Fig. 5.1 which can lead to isomerization and racemization. Appropriate choice of ligands and co-catalysts (promoters) may enable these difficulties to be overcome.

New catalyst systems which operate at even milder reaction conditions will undoubtedly be discovered and developed but their likely commercialization can only be the outcome of detailed economic assessments. As indicated previously there is a lower limit to the operating pressure which could be tolerated in a commercial process, and this is determined by chemical engineering constraints. The development of satisfactory single reactor processes to transform olefins directly into alcohols, e.g. propylene to 2-ethylhexanol, remains a desirable target.

The hydroformylation reaction forms only part of a general scheme for the reactions of olefins with carbon monoxide and other co-reactants,[17] (reaction (4)).

$$RCH = CH_2 + CO + R'OH \rightarrow RCH_2CH_2CO_2R' \qquad (4)$$

If $R' = H$, carboxylic acids are the products whereas if $R' =$ alkyl then carboxylic acid esters are produced. These reactions may represent future opportunities. BASF recently announced the construction of the first plant to utilize this chemistry, for the production of adipic acid from butadiene.[18] This 60 000 tonne year^{-1} process is to involve a two-stage carbomethoxylation of butadiene, using carbon monoxide and methanol to produce the dimethyl ester of adipic acid which is subsequently converted into the parent acid by hydrolysis (Fig. 5.2). Adipic acid is an intermediate for nylon 6,6 fibres (see also section 5) and resins. Although the nature of the catalysts have not been disclosed they probably contain cobalt together with a nitrogen-containing base in the second stage to promote isomerization from internal to terminal unsaturation with the resultant formation of the linear di-ester.

Finally, there has been considerable interest over the years in the search for surface-supported hydroformylation catalysts, i.e. heterogeneous analogues of the homogeneous catalysts, with the aims of facilitating the separation of catalyst from reactants and products and of eliminating the problems of catalyst deposition onto the walls of the reactor. These attempts appear to have been unsuccessful, in spite of claims to the contrary, because of slow leaching and consequent loss of the catalyst from the support into the liquid phase.[19]

$$CH_2{=}CH{-}CH{=}CH_2 + CO + CH_3OH \rightarrow CH_3CH{=}CHCH_2CO_2CH_3$$

$$CH_3CH{=}CHCH_2CO_2CH_3 + CO + CH_3OH \rightarrow CH_3CO_2CH_2CH_2CH_2CH_2CO_2CH_3$$

$$\begin{matrix} CH_2CH_2CO_2CH_3 \\ | \\ CH_2CH_2CO_2CH_3 \end{matrix} \quad +2H_2O \rightarrow \quad \begin{matrix} CH_2CH_2CO_2H \\ | \\ CH_2CH_2CO_2H \end{matrix} \quad +2CH_3OH$$

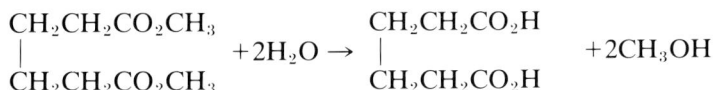

Fig. 5.2. Carbonylation of butadiene to adipic acid.

4 Production of acetic acid and the acetyl chemicals

Acetic acid has been manufactured in large quantities for over 100 years. It is currently produced on a scale of c. 3 million tonnes per annum worldwide, and its principal uses are as an intermediate in the manufacture of cellulose acetate (via acetic anhydride), vinyl acetate monomer (the polymer is used in emulsion paints) and acetate esters (solvents). The changes in the processes used for the production of acetic acid (most of which involve catalysis, or at least initiation, by transition metal ions or complexes) during this 100-year period reflect some of the underlying general trends in the chemical industry, in particular towards more energy efficient processes and cheaper feedstocks.

Acetic acid was initially produced by fermentation, as were many other organic chemicals in the early days of the chemical industry. The first major synthetic process was based on the hydration or hydrolysis of acetylene to acetaldehyde (the parent molecule of the acetyl chemicals family) catalysed by mercuric ion, followed by oxidation to either acetic acid or acetic anhydride (reaction (5)).

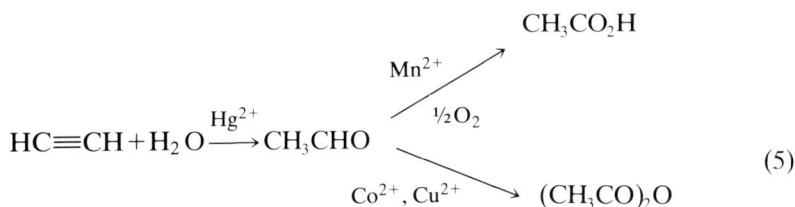

$$HC{\equiv}CH + H_2O \xrightarrow{Hg^{2+}} CH_3CHO \begin{array}{c} \nearrow^{Mn^{2+}, \frac{1}{2}O_2} CH_3CO_2H \\ \searrow_{Co^{2+}, Cu^{2+}} (CH_3CO)_2O \end{array} \tag{5}$$

This process was used extensively in Europe, with the alternative oxidation of ethanol catalysed by cobalt and chromium acetates predominating in the USA. In the latter process metal ions are involved in the initiation step, namely radical generation, but the oxidation reaction itself is believed to involve a radical chain mechanism. These routes were the dominant processes utilized for the production of acetic acid for over 40 years until the period 1955–60 when two new developments occurred.

The first of these involved the discovery of two alkane oxidation processes, namely the short chain, e.g. butane, oxidation developed by Celanese in the USA and the naphtha oxidation process discovered by BP in Europe. In the former a metal salt, usually cobalt acetate, is used as an initiator to direct the cleavage of the alkane chain to give maximum yields of acetic acid (reaction (6)).

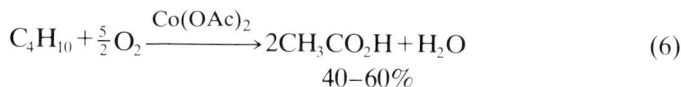

$$C_4H_{10} + \tfrac{5}{2}O_2 \xrightarrow{\text{Co(OAc)}_2} 2CH_3CO_2H + H_2O \qquad (6)$$
$$40\text{--}60\%$$

This process is of relatively low selectivity and the economics are therefore heavily dependent upon the sale of by-products. Alkane oxidation is a radical chain process and the main propagation steps need not involve the metal ion. Cobalt is thought to participate in the initiation and decomposition of alkyl hydroperoxide intermediates.

The second development was the discovery by Wacker Chemie of a simple, high yield oxidation of ethylene to acetaldehyde or, in the presence of acetic acid, to vinyl acetate (reaction (7)).

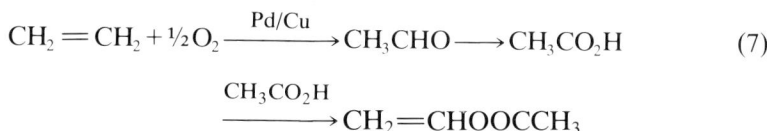

$$CH_2=CH_2 + \tfrac{1}{2}O_2 \xrightarrow{\text{Pd/Cu}} CH_3CHO \longrightarrow CH_3CO_2H \qquad (7)$$
$$\xrightarrow{\text{CH}_3\text{CO}_2\text{H}} CH_2=CHOOCCH_3$$

A further development which now completely dominates all new processes for acetic acid production is that based on the carbonylation of methanol, a homogeneously-catalysed process which was initially commercialized by BASF in 1966 (using cobalt catalysts) and a much lower pressure rhodium-catalysed process pioneered by Monsanto in 1971. Today the Monsanto technology accounts for roughly one-third of acetic acid production.

The overall trends in the production of acetic acid and acetyl chemicals reflect a shift away from high energy (and usually expensive) intermediates such as acetylene, towards lower energy materials (and successively more economic feedstocks) such as paraffins, olefins, methanol and synthesis gas. The latest development in this area is the commercialization by Tennessee Eastman/Halcon of a process for the carbonylation of methyl acetate to acetic anhydride, thus completing the first totally synthesis gas-based route to this molecule. Many of these developments have parallels with discoveries in the use of soluble metal complexes as catalysts which are amplified by the following discussion of (i) the oxidation of ethylene to acetaldehyde, (ii) the carbonylation of methanol to acetic acid, and (iii) the carbonylation of methyl acetate to acetic anhydride.

4.1   Oxidation of ethylene to acetaldehyde

It had been known since 1894 that ethylene could be oxidized stoichiometrically to acetaldehyde,[20] by aqueous palladium chloride, with the formation of HCl and metallic palladium (reaction (8)).

$$CH_2{=}CH_2 + [PdCl_4]^{2-} + H_2O \rightarrow CH_3CHO + Pd^0 + 2HCl + 2Cl^- \qquad (8)$$

In 1956, Wacker Chemie discovered that the reaction could be made catalytic by (a) re-oxidizing the palladium with cupric chloride (reaction (9)) and (b) re-oxidizing the reduced cuprous ion with oxygen (reaction (10))

$$2Cl^- + Pd^0 + 2CuCl_2 \rightarrow [PdCl_4]^{2-} + 2CuCl \qquad (9)$$

$$2CuCl + \tfrac{1}{2}O_2 + 2HCl \rightarrow 2CuCl_2 + H_2O \qquad (10)$$

thus giving the overall reaction (11).

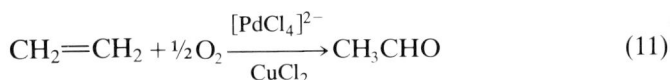

$$CH_2{=}CH_2 + \tfrac{1}{2}O_2 \xrightarrow[\text{CuCl}_2]{[\text{PdCl}_4]^{2-}} CH_3CHO \qquad (11)$$

At the chloride concentrations used in the Wacker process, copper is probably present in solution as $CuCl_2$ and CuCl rather than as chlorocuprate ions. Acetaldehyde is produced in yields of c. 95%. Such was the significance of this reaction that it achieved commercial status only 4 years after its discovery. This was not only the first industrial catalytic oxidation reaction to use transition metal–olefin chemistry but also the first demonstration that a precious metal could be used economically in a homogeneously-catalysed reaction on an industrial scale.[21]

The original Wacker process operates in two stages. In the first step ethylene and oxygen, in appropriate proportions to ensure almost complete conversion of ethylene, are introduced at a total pressure of 10 atm to a solution of cupric chloride and palladium chloride in dilute HCl. This solution is continuously circulated through a second reactor where catalyst regeneration by oxidation with air at 10 atm and 100–110°C takes place. In the one step modification introduced by Hoechst the catalyst is regenerated *in situ*. Ethylene and oxygen at 3 atm are passed into the reactor containing a solution of catalyst maintained at 100°C. The composition of the gas mixture is kept well above the upper flammability limit so ethylene is present in large excess. Conditions are adjusted to give about 40% ethylene conversion and total oxygen consumption. The heat produced by the reaction is used to distil off the acetaldehyde and unreacted ethylene is recycled.

Although some details of the reaction mechanism are unclear[22] the generally accepted details are summarized in Fig. 5.3.

The process remains the most elegant application of homogeneous catalytic oxidation and attempts to develop a heterogeneous version have been unsuccessful. One reason why the homogeneous process has maintained its commercial competitiveness is that the product is volatile and therefore easily separated from

$$[PdCl_4]^{2-} + C_2H_4 \rightleftharpoons [(C_2H_4)PdCl_3]^- + Cl^-$$

$$[(C_2H_4)PdCl_3]^- + H_2O \longrightarrow \left[\begin{array}{c} CH_2{=}CH_2 \\ Cl \\ Pd \\ Cl \quad OH_2 \end{array}\right] \rightleftharpoons \left[\begin{array}{c} CH_2{=}CH_2 \\ Cl \\ Pd \\ H_2O \quad Cl \end{array}\right] + Cl^-$$

$$\left[\begin{array}{c} CH_2{=}CH_2 \\ Cl \\ Pd \\ H_2O \quad Cl \end{array}\right] \rightleftharpoons \left[\begin{array}{c} CH_2{=}CH_2 \\ Cl \\ Pd \\ HO \quad Cl \end{array}\right]^- + H^+$$

$$\left[\begin{array}{c} CH_2{=}CH_2 \\ Cl \\ Pd \\ HO \quad Cl \end{array}\right]^- + H_2O \longrightarrow \left[\begin{array}{c} HOCH_2CH_2 \quad Cl \\ Pd \\ H_2O \quad Cl \end{array}\right]^-$$

$$\left[\begin{array}{c} HOCH_2CH_2 \quad Cl \\ Pd \\ H_2O \quad Cl \end{array}\right]^- \longrightarrow CH_3CHO + Pd^0 + H^+ + H_2O + 2Cl^-$$

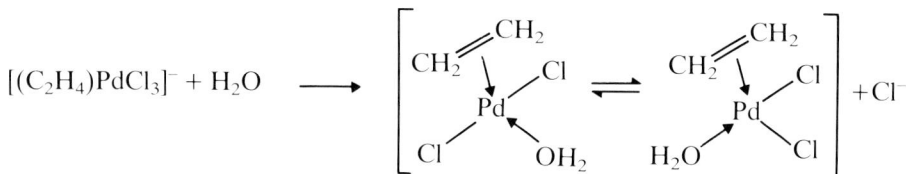

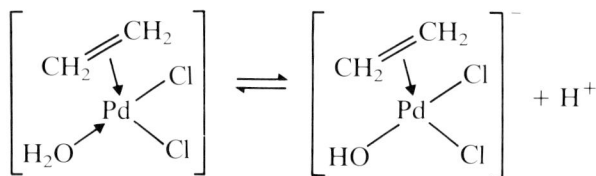

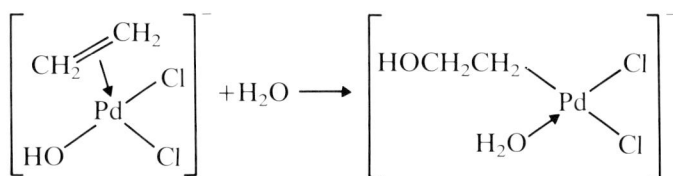

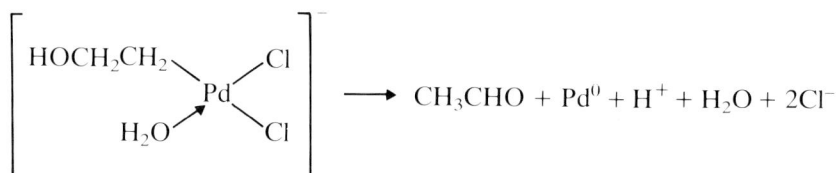

Fig. 5.3. Reaction mechanism for the oxidation of ethylene to acetaldehyde.

the catalyst. In 1968, immediately before the development of the Monsanto process for methanol carbonylation, 100% Japanese, 46% European and 39% US acetaldehyde production was based on ethylene.

An extension to this process is the formation of vinyl acetate, where the oxidation reaction is carried out in acetic acid rather than water. Thus, in 1960 it was discovered that vinyl acetate could be produced by passing ethylene into acetic acid solutions of palladium chloride containing sodium or lithium acetate at 10 atm total pressure and 120–130°C (reaction (12)).

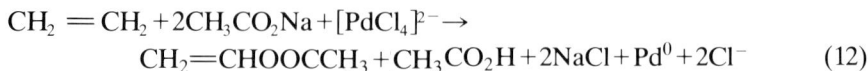

$$CH_2{=}CH_2 + 2CH_3CO_2Na + [PdCl_4]^{2-} \rightarrow$$
$$CH_2{=}CHOOCCH_3 + CH_3CO_2H + 2NaCl + Pd^0 + 2Cl^- \qquad (12)$$

Reaction (12) together with (9) and (10) give the net reaction (13).

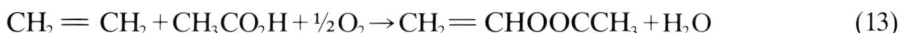

$$CH_2{=}CH_2 + CH_3CO_2H + \tfrac{1}{2}O_2 \rightarrow CH_2{=}CHOOCCH_3 + H_2O \qquad (13)$$

Oxygen is used for the *in situ* regeneration of the cupric ion which in turn re-oxidizes zerovalent palladium. Overall yields of 90 and 95% based on ethylene and acetic acid respectively were claimed. This homogeneously-catalysed process was first commercialized, but later abandoned, in favour of a heterogeneous gas phase counterpart based upon the same chemical principle (reaction (14)) because of severe corrosion problems.

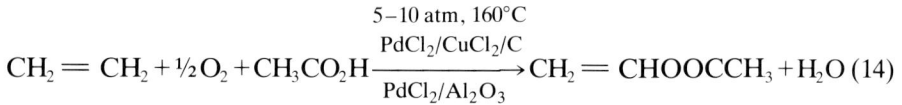

$$CH_2 = CH_2 + \tfrac{1}{2}O_2 + CH_3CO_2H \xrightarrow[\substack{PdCl_2/Al_2O_3}]{\substack{5-10 \text{ atm, } 160°C \\ PdCl_2/CuCl_2/C}} CH_2 = CHOOCCH_3 + H_2O \quad (14)$$

Again, greater than 90% selectivities to vinyl acetate are achieved.

4.2   Carbonylation of methanol to acetic acid

In 1966 BASF described a high pressure process based on Reppe chemistry for the carbonylation of methanol to acetic acid using an iodide-promoted cobalt catalyst.[23] This was rapidly followed in 1968 by Monsanto's announcement of the discovery of a low pressure carbonylation using an iodide-promoted rhodium catalyst,[24] a process which was commercialized in 1971.[25] A comparison between the salient features of the two processes is given in Table 5.7, which demonstrates similar trends and advantages associated with the substitution of rhodium for cobalt in hydroformylation as discussed previously. Thus, the lower activity of cobalt, relative to rhodium, requires higher temperatures and as a consequence very high pressures of carbon monoxide are necessary to stabilize the cobalt carbonyl at these temperatures. Not only is the selectivity of the high pressure process lower, as a consequence of greater by-product formation, but it is also sensitive to traces of hydrogen (as impurities in the CO feed), whereas it is claimed that the presence of up to 50% hydrogen is not detrimental to the Monsanto system.

Table 5.7. Comparison of methanol carbonylation processes based on cobalt and rhodium

	Cobalt (BASF)	Rhodium (Monsanto)
Pressure (atm)	500–700	30–40
Temperature (°C)	230	180
Metal concentration	$\sim 10^{-1} M$	$\sim 10^{-3} M$
Promoter	I^-	I^-
Selectivity (%) (based on methanol)	~ 90	> 99
By-products	$CH_4, CH_3CHO, C_2H_5OH, CO_2$	None (detection limit 0.1%)

Table 5.8. Kinetics of methanol carbonylation processes based on cobalt and rhodium

Reaction variable	Effect on reaction rate	
	Cobalt	Rhodium
$[CH_3OH]$	First order	Zero order
P_{CO}	Second order	Zero order
$[I^-]$	First order	First order
$[M]$	Variable	First order
Products $[CH_3CO_2H]$ and $[CH_3CO_2CH_3]$	—	Zero order

A comparison between the kinetic behaviour of the two processes (Table 5.8) highlights significant differences which are reflected in different rate-determining steps. The cobalt-catalysed reaction is dependent upon both methanol concentration and the partial pressure of carbon monoxide and the mechanism, illustrated in Fig. 5.4, is thought to be analogous to that of the cobalt-catalysed hydroformylation reaction. In contrast, kinetic studies on the rhodium-catalysed methanol carbonylation reaction show a remarkably simple behaviour in which the rate is independent of the concentrations of both reactants and products. Neither therefore have any kinetic influence. Interestingly, the kinetics of a heterogenously-catalysed rhodium system, developed in parallel with the homogeneous process, are very similar.[26] However, the heterogeneous analogue was characterized by significantly reduced activity and slow loss of precious metal from the support. Thus, the homogeneous process was preferred.

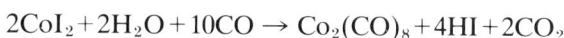

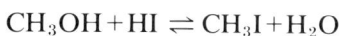

$$2CoI_2 + 2H_2O + 10CO \rightarrow Co_2(CO)_8 + 4HI + 2CO_2$$

$$Co_2(CO)_8 + H_2O + CO \rightarrow 2HCo(CO)_4 + CO_2$$

$$CH_3OH + HI \rightleftharpoons CH_3I + H_2O$$

$$CH_3I + HCo(CO)_4 \rightarrow HI + CH_3Co(CO)_4$$

$$CH_3Co(CO)_4 \rightleftharpoons CH_3COCo(CO)_3 \overset{+CO}{\underset{-CO}{\rightleftharpoons}} CH_3COCo(CO)_4$$

$$CH_3COCo(CO)_4 + H_2O \rightarrow CH_3CO_2H + HCo(CO)_4$$

Fig. 5.4. Reaction mechanism for the cobalt-catalysed carbonylation of methanol to acetic acid.

A key intermediate in this system is believed to be $[Rh(CO)_2I_2]^-$ which is the principal species present in solution at 6 atm pressure and 100°C according to high pressure infra-red spectroscopic measurements.[27] The rate-determining step in the catalytic cycle is the oxidative addition of methyl iodide to $[Rh(CO)_2I_2]^-$ to give a methyl rhodium complex (not detected spectroscopically) which is very rapidly transformed into an acetyl species under the pressures of carbon monoxide used. A binuclear rhodium acetyl iodo complex $[CH_3CORh(CO)I_3]_2^{2-}$,

$$CH_3OH + HI \rightleftharpoons CH_3I + H_2O$$

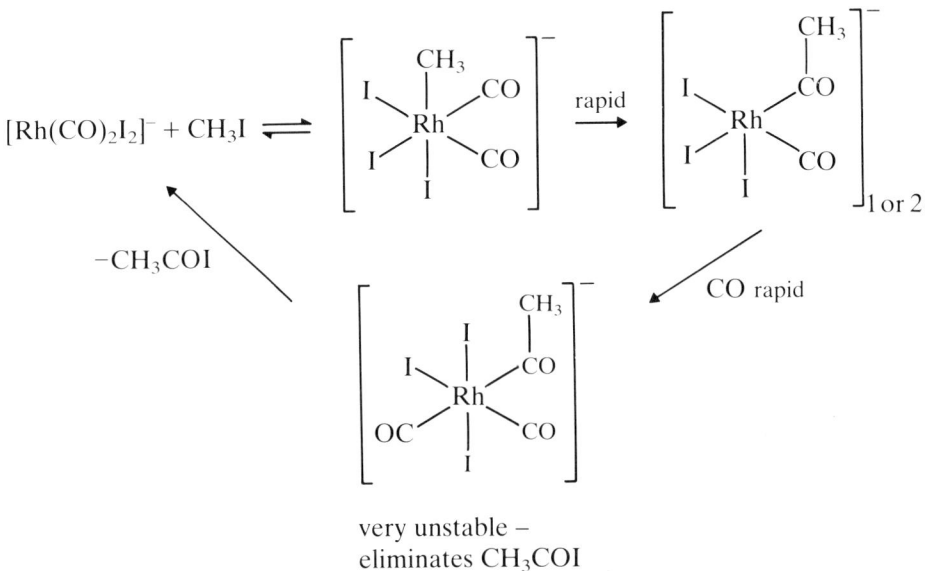

Fig. 5.5. Reaction mechanism for the rhodium-catalysed carbonylation of methanol to acetic acid.

dimerized through a very weak iodide bridge, has been isolated from these reactions and structurally characterized. The overall catalytic cycle is believed to operate as summarized in Fig. 5.5. The key to the high selectivity of the process is probably associated with the position of the rate-determining step and the ease of formation of the acyl species. The instability/short lifetime of the methyl rhodium complex under reaction conditions makes it unlikely that a reaction of the type (15) would be a serious competing reaction.

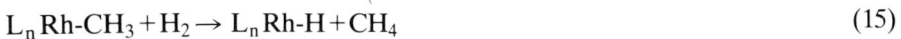

$$L_n Rh\text{-}CH_3 + H_2 \rightarrow L_n Rh\text{-}H + CH_4 \tag{15}$$

Overall, therefore, in spite of the use of a precious metal catalyst, the Monsanto process has significant operating advantages over the BASF process and is replacing this technology. In 1972 the Monsanto process accounted for approximately 10% of the total capacity of acetic acid production in the USA whereas in 1982 this had risen to 40%. The worldwide production of acetic acid is of the order of 3 million tonnes per annum and when plants using Monsanto technology which are currently under construction are completed this will represent about one-third, i.e. 1 million tonnes per annum.

Iridium has also been shown to be an active catalyst for methanol carbonylation but it is not as selective as rhodium. The lower selectivity appears to be associated with a more complicated catalytic cycle.[25] Other non-precious metals have also been examined for their potential in methanol carbonylation. For example, high selectivities to acetic acid have been observed over nickel catalysts, but side reactions such as hydrocarbon formation also appear significant.

4.3   Carbonylation of methyl acetate to acetic anhydride

Although the cobalt halide-catalysed carbonylation of methyl acetate to acetic anhydride was discovered by Reppe in the 1950s, the reaction required very high pressures and gave only low reaction rates. Consequently it did not achieve commercial status. In what may be considered to be an extension of the previously described Monsanto technology, the Tennessee Eastman-Halcon process[28] for the carbonylation of methyl acetate to acetic anhydride comprises the most recent application of homogeneous catalysis. This 227 000 tonne year^{-1} capacity process, which came on stream at Kingsport, Tennessee in October 1983, is an example of an integrated synthesis gas system based entirely on coal as feedstock (reaction (16)).

$$\text{Coal} \rightarrow \text{CO/H}_2 \rightarrow \text{CH}_3\text{OH} \xrightarrow{\text{CH}_3\text{CO}_2\text{H}} \text{CH}_3\text{CO}_2\text{CH}_3 \xrightarrow{\text{CO}} (\text{CH}_3\text{CO})_2\text{O}_2 \qquad (16)$$

Good quality coal, available locally, is gasified to CO/H_2 which in turn is converted into methanol. The process then operates in two stages, namely the reaction of methanol with acetic acid to give methyl acetate, followed by carbonylation to acetic anhydride. Both processes are highly selective and the overall selectivity is said to be in excess of 95%, based on methanol. The acetic anhydride produced is used internally and converted, with cellulose, into cellulose acetate for applications as a photographic film base, textile fibres and cigarette filters. A by-product of this esterification is acetic acid which is used to esterify the methanol which is produced separately from CO/H_2. Acetic anhydride is conventionally manufactured either by the energy intensive and relatively inefficient ketene route or by the oxidation of acetaldehyde.

Few details of the chemistry and reaction conditions are available but the process is believed to operate at relatively low pressures (30 atm) and modest temperatures (170–200°C) comparable with the Monsanto methanol to acetic acid technology. The carbonylation step is believed to involve a multicomponent rhodium-based homogeneous catalyst containing iodide together with promoters such as pyridine bases and chromium or lithium salts. The chemistry is probably closely related to that of methanol carbonylation, with the rhodium-catalysed carbonylation of methyl iodide to acetyl iodide being the key step (reactions (17)–(20)).

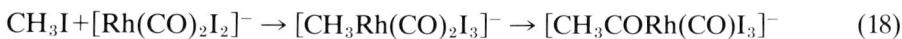

$$CH_3CO_2CH_3 + I^- \rightarrow CH_3I + CH_3CO_2^- \tag{17}$$

$$CH_3I + [Rh(CO)_2I_2]^- \rightarrow [CH_3Rh(CO)_2I_3]^- \rightarrow [CH_3CORh(CO)I_3]^- \tag{18}$$

$$[CH_3CORh(CO)I_3]^- + CO \rightarrow CH_3COI + [Rh(CO)_2I_2]^- \tag{19}$$

$$CH_3COI + CH_3CO_2^- \rightarrow (CH_3CO)_2O + I^- \tag{20}$$

The rhodium–iodide catalysed reaction is rather slower than methanol carbonylation unless certain promoters are added. Typical promoters are pyridines, phosphines and other materials which will form ionic quaternary iodides in the presence of methyl iodide. These quaternary iodides are believed to catalyse reaction (20). This has the beneficial effect of decreasing the steady-state concentration of acetyl iodide, the presence of which can partially poison the rhodium catalyst through the formation of molecular iodine. In the presence of these co-catalysts the rates of methyl acetate carbonylation can approach those of methanol carbonylation.

A kinetic and spectroscopic investigation[29–31] on a catalyst system comprising rhodium trichloride hydrate, methyl iodide, an organic base (pyridine, triphenylphosphine) and chromium compounds has demonstrated that the reaction rate is independent both of carbon monoxide pressure above 15 atm and of the methyl acetate concentration. The reaction is first order with respect to rhodium, methyl iodide and to the base at low concentrations. Chromium compounds have less influence on the reaction rate but considerably reduce the induction period, presumably by assisting the catalyst activation process. Oxidative addition of methyl iodide followed by rapid insertion of carbon monoxide is believed to be the rate-determining step in the catalytic cycle, similar to the carbonylation of methanol.

Certain economic factors particular to Tennessee Eastman, e.g. substantial local deposits of high grade coal, internal use of product acetic anhydride, recycling of acetic acid, etc., make the process particularly favourable to these operators, but nevertheless the production of acetic anhydride by the carbonylation of methyl acetate does provide another example of the general trend away from oil-based feedstocks to coal and/or natural gas. As a possible extension of this technology Halcon, amongst others, has also been active in the development of homogeneously-catalysed synthesis gas based routes for the production of other acetyl chemicals. For example, the carbonylation of methyl acetate to ethylidene diacetate, in the presence of significant pressures of hydrogen, followed by thermal cracking to vinyl acetate and acetic acid has been investigated.[32] The co-produced acetic acid may be recycled by reaction with methanol to produce the starting acetate (reactions (21)–(23))

$$2CH_3CO_2CH_3 + 2CO + H_2 \rightarrow CH_3CH(OCOCH_3)_2 + CH_3CO_2H \qquad (21)$$

$$CH_3CH(OCOCH_3)_2 \rightarrow CH_2 = CHOOCCH_3 + CH_3CO_2H \qquad (22)$$

$$2CH_3OH + 2CH_3CO_2H \rightarrow 2CH_3CO_2CH_3 + 2H_2O \qquad (23)$$

giving the overall reaction (24).

$$2CH_3OH + 2CO + H_2 \rightarrow CH_2 = CHOOCCH_3 + 2H_2O \qquad (24)$$

Both rhodium and palladium have been claimed as good catalysts for this reaction. However, the system appears to be at a considerably less advanced stage of development than the carbonylation of methyl acetate to acetic anhydride.

5   Hydrocyanation of butadiene

Adiponitrile is a key intermediate in the production of nylon 6, 6 through its reduction to hexamethylenediamine. The direct hydrocyanation of butadiene to adiponitrile, discovered by DuPont,[33] is based on the nickel-catalysed double addition of HCN to butadiene (reaction (25))

$$CH_2 = CH-CH = CH_2 + 2HCN \rightarrow NC(CH_2)_4CN \qquad (25)$$

which is now operated at three locations: Orange (Texas), Victoria (Texas)—the latter a conversion in 1983 from a plant which used the indirect hydrocyanation route via dichlorobutene—and in France (in a Rhone Poulenc/DuPont joint venture), representing a total capacity of 458000 tonnes year^{-1}. As indicated this new process is tending to replace the old three-stage indirect hydrocyanation of butadiene which was previously developed by DuPont.

The direct hydrocyanation process has been in operation since 1971 few details on process conditions and reaction mechanism are available. This homogeneously-catalysed reaction is carried out in three stages, each of which is operated under relatively mild conditions of temperature and pressure. There are no known heterogeneous catalysts that will accomplish these transformations satisfactorily.

The first stage, reaction (26), involves the addition of HCN to butadiene catalysed by a nickel–aryl phosphite complex in the presence of excess ligand, and produces an approximately 2:1 mixture of the linear and branched mononitriles, 3-pentenenitrile (containing a minor amount of 4-pentenenitrile) and 2-methyl-3-butenenitrile, respectively. This reaction is operated at a modest pressure sufficiently above ambient to ensure that unreacted butadiene is condensed.

Subsequent stages are operated at atmospheric pressure.

$$CH_2=CH-CH=CH_2+HCN \xrightarrow[100°C]{NiL_4/L} \overset{70\%}{CH_3CH=CHCH_2CN} + \overset{30\%}{CH_2=CHCHCH_3}$$

$$L = (ArO)_3P$$

$$\underset{70\%}{} \quad \underset{30\%}{} \quad (26)$$

In the second stage the mixed product is isomerized into a mixture of 3- and 4-pentenenitriles using a similar catalyst system in the presence of a Lewis acid promoter such as zinc chloride (reaction (27)).

$$CH_2=CHCHCH_3 \underset{\xleftarrow{} 120°C}{\overset{NiL_4/ZnCl_2}{\rightleftharpoons}} CH_2=CHCH_2CH_2CN+CH_3CH=CHCH_2CN$$

$$\underset{|}{}$$
$$CN \qquad\qquad\qquad\qquad\qquad\qquad\qquad\qquad (27)$$

At this stage it is necessary to avoid the formation of other isomeric products, since the most stable conjugated nitrile, $CH_3CH_2CH=CHCN$, is thought to be a catalyst inhibitor in the final step. The third and final stage comprises the addition of HCN to the mixture of 3- and 4-pentenenitriles using a similar catalyst system (reaction (28)).

$$\left.\begin{array}{c} CH_2=CHCH_2CH_2CN \\ + \\ CH_3CH=CHCH_2CN \end{array}\right\} \begin{array}{c} NiL_4/L \\ +HCN \quad\rightarrow \\ ZnCl_2 \\ 80°C \end{array} \overset{83\%}{NC(CH_2)_4CN} \qquad (28)$$

The major by-products are 2-methylglutaronitrile and a small amount of ethyl-succinonitrile. The use of zinc chloride as the Lewis acid gives adiponitrile in c. 83% selectivity at 80°C. It is believed that replacement of $ZnCl_2$ by triphenyl-boron gives a significant improvement in selectivity (to > 90% at 99% conversion in the third stage) at lower operating temperatures (40–50°C). The role of the ligand is crucial. Among phosphorus compounds, only phosphites produce active catalysts and among phosphites only the relatively bulky triarylphosphites, e.g. $P(O\text{-}o\text{-tolyl})_3$, give high selectivity in the final stage. Presumably the steric effect of the bulky ligands also directs the reaction in favour of the formation of linear rather than branched products.

Mechanistically the first stage can be thought of in terms of the sequence of reactions outlined in Fig. 5.6, namely oxidative addition of HCN to the nickel complex (step 1), coordination of butadiene (step 2) and formation of a π-allyl intermediate through a cis-rearrangement of the nickel hydride (step 3) followed by reductive elimination of the mononitrile (step 4) and regeneration of the

$$NiL_4 + HCN \rightleftharpoons HNiL_4(CN) \rightleftharpoons HNiL_3(CN) + L \qquad (1)$$

$$HNiL_3(CN) + CH_2{=}CH{-}CH{=}CH_2 \rightarrow HNiL_3(CN) \qquad (2)$$
$$\uparrow$$
$$CH_2{=}CH{=}CH_2$$

$$HNiL_3(CN) \quad \rightarrow \quad H{-}\overset{CH_2}{\underset{CH}{C}}{-}NiL_3(CN) \qquad (3)$$
$$\uparrow$$
$$CH_2{=}CHCH{=}CH_2 \qquad\qquad CH_3$$

$$H{-}\overset{CH_2}{\underset{CH}{C}}{-}NiL_3(CN) \rightarrow CH_3CH{=}CH{-}CH_2CN + NiL_3 \qquad (4)$$
$$CH_3$$

$$NiL_3 + HCN \rightleftharpoons HNiL_3(CN) \qquad (5)$$

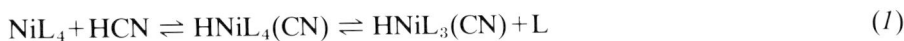

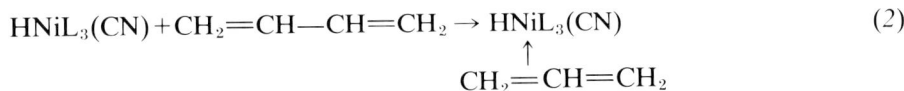

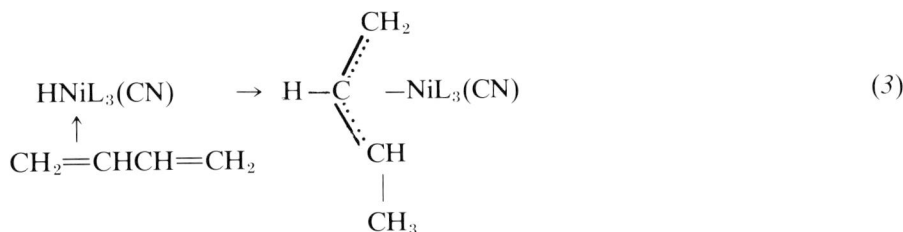

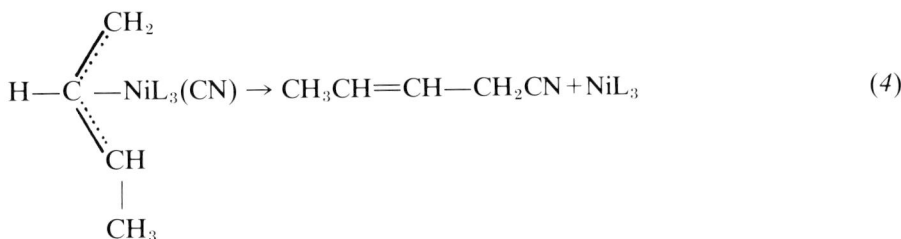

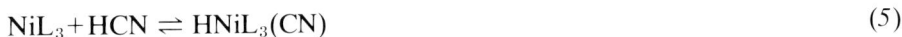

Fig. 5.6. Reaction mechanism of the first stage in the hydrocyanation of butadiene.

catalyst by addition of HCN (step 5). Recent mechanistic studies[34-36] on the addition of HCN to olefins using Ni[P(O-o-tolyl)$_3$]$_3$ as catalyst precursor are broadly consistent with this scheme. The RNiL$_m$CN reaction intermediates have been identified spectroscopically in some cases.

The effect of the Lewis acid co-catalyst in the second and third stages of the process seems to be to ensure the rapid isomerization of the pentene- and butene-nitriles to 4-pentenenitrile and to facilitate the addition of HCN to the non-conjugated double bond in an anti-Markownikoff manner. Spectroscopic studies have confirmed that the Lewis acids can coordinate strongly to the nitrogen lone electron pair of the hydrido-cyanide intermediates to form adducts of the HNiL$_3$CN.A type. It has been proposed that the intriguing effect of the Lewis acids on hydrocyanation involves increasing the effective concentration of nickel in the catalytic loop, accelerating the rate of carbon–carbon coupling to form alkenenitriles from alkyl nickel cyanide complexes and destabilizing relatively bulky branched alkyl intermediates relative to less crowded linear ones.

6   Shell higher olefins process

Although the majority of olefin oligomerizations involve alkyl aluminium growth reactions, a relatively recent development which involves at least one homogeneous transition metal-catalysed stage is the Shell higher olefins process (SHOP). SHOP is a process which was discovered in 1966 for converting ethylene into detergent range olefins (actually C_{10}-C_{18} even numbered linear α-olefins) in high yield using a combination of oligomerization, isomerization and metathesis reactions. It was first commercialized in 1977 when a 115 000 tonnes year^{-1} capacity plant came into operation at Geismar, Louisiana[37] and a second plant (117 000 tonnes year^{-1} capacity) was brought on stream at Stanlow, UK, in 1981.

In the first step of this process (reaction (29)) ethylene is selectively oligomerized to give, in a geometric Schulz–Flory distribution, even carbon number linear (99+%) α-olefins (93–99%) using a nickel-based homogeneous catalyst system obtained from the reaction of bis(cyclooctadienyl) nickel with a substituted phosphine of the type $R_2PCH_2COO^-$, where R = phenyl or cyclohexyl. Quoted catalytic activities are of the order of 0.6 mol ethylene mol^{-1} Ni s^{-1}.[38]

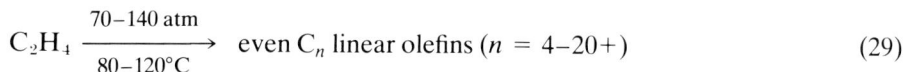

$$C_2H_4 \xrightarrow[80-120°C]{70-140\ atm} \text{even } C_n \text{ linear olefins } (n = 4\text{--}20+) \qquad (29)$$

The reaction is carried out in a polar solvent such as 1,4-butanediol which is largely immiscible with the α-olefin product. The chelating $\widehat{PO}$ functionality of the phosphines is an essential feature of the catalyst which gives excellent activity and ensures solubility in the polar solvent. The high activity of the catalyst enables low metal concentrations to be used. The reaction mixture comprises three phases: (i) solvent containing catalyst, (ii) oligomer product and (iii) ethylene gas. As hydrocarbon product is formed it separates from the solvent–catalyst phase and enters the hydrocarbon phase. After the reaction the ethylene is stripped and the oligomer product phase separates from the catalyst/solvent phase. Two principal advantages arise from this method of operation. First, the catalyst phase may be directly recycled to the oligomerization reactor and, secondly, the formation of branched olefins by secondary reactions is minimized because of the very low concentrations of α-olefins in the catalyst/solvent phase.

High partial pressures of ethylene are required for good rates of reaction and high linearity of the α-olefins product. The latter is characterized by a geometric growth factor K which is defined as in (30)

$$K = \frac{\text{moles } C_{n+2} \text{ olefin}}{\text{moles } C_n \text{ olefin}} \qquad (30)$$

Control of K is the key to the process since it sets the product distribution of the α-olefin production and also determines the average carbon number of the overall process (see Fig. 5.7). K may be adjusted by varying the catalyst composition. Products are formed in the C_n range $n = 4$–$20+$, many of which are outside the detergent range and of low commercial value. The C_{10}-C_{18} α-olefin products, which in fact form less than 30% of the total product, are separated for sale as high purity α-olefins. The remainder, butene, C_6, C_8 and C_{20+} fractions, which are all of much lower commercial value, are then fed to the second stage.

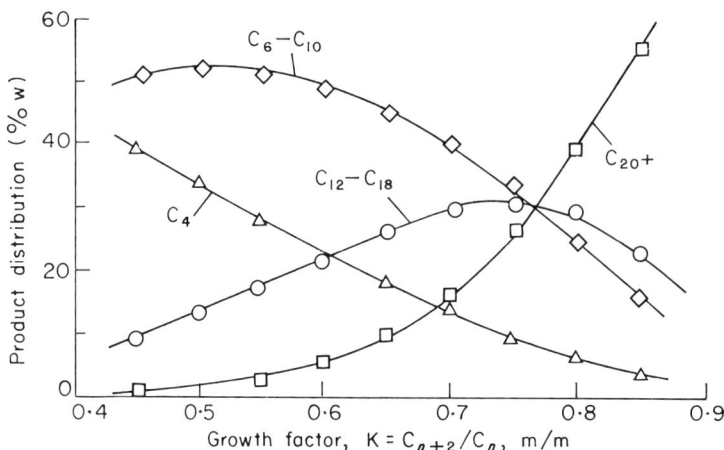

Fig. 5.7. Oligomer product distribution vs. K factor.

The second stage of the process (reaction (31)) comprises an isomerization step in which the aforementioned olefins are isomerized to a near-equilibrium distribution of straight chain internal olefins over an undisclosed, probably heterogeneous, catalyst under the reaction conditions indicated.

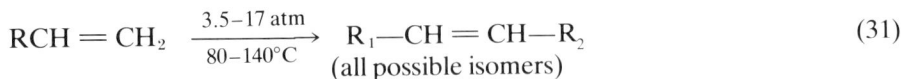

$$RCH = CH_2 \xrightarrow[80-140°C]{3.5-17\ atm} R_1—CH = CH—R_2 \qquad (31)$$
$$\text{(all possible isomers)}$$

The reaction is believed to occur by a stepwise mechanism, and the double bond is shifted from the terminal position into internal positions approaching equilibrium. The equilibrium is characterized by a low concentration of the double bond in the terminal (α) position and an almost statistical distribution over the other positions of the molecule. Present at this stage of the process is a mixture of internal olefins in two discrete carbon number fractions, namely C_4-C_8 (larger fraction) and C_{20+}, which is then fed to the next step of the process.

The third stage, which probably also involves a heterogeneous catalyst such as MoO_3/Al_2O_3, is the metathesis, or disproportionation, step in which a statistical redistribution of the olefins occurs about the double bonds, e.g. for the production of a C_{12} internal olefin (reaction (32)).

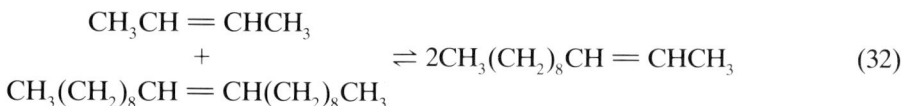

$$CH_3CH = CHCH_3$$
$$+ \qquad\qquad \rightleftharpoons 2CH_3(CH_2)_8CH = CHCH_3 \qquad (32)$$
$$CH_3(CH_2)_8CH = CH(CH_2)_8CH_3$$

This process yields about 10–15 wt % of the desired detergent range olefins per pass. The products may then be separated by distillation as C_{11}-C_{14} linear olefins with a typical composition indicated in Table 5.9.

Table 5.9. Typical composition of SHOP isomerization/metathesis products (wt %)

Olefin carbon number	C_{11}/C_{12}	C_{13}/C_{14}
10	0.5	—
11	54	—
12	45	< 1.0
13	< 1.0	55
14	—	44
15	—	< 0.5

Of these products > 96% comprise linear olefin. Since the metathesis stage only yields 10–15 wt % of the desired detergent range product extensive recycle of the lower (< C_{10}) and higher (> C_{15}) carbon number materials is required. These are separated by distillation and the < C_{10} product recycled to metathesis and the > C_{15} product returned through the isomerization and metathesis stages. The product olefins can of course be converted into detergent alcohols by hydroformylation and hydrogenation.

The Shell higher olefins process represents an elegant example of how one can work within the limitations of the Schulz-Flory product distribution from a polymerization type reaction yet still optimize on the higher value products by using the sequence of reactions described. SHOP thus solves the basic problem of a mis-match between product distribution from the oligomerization process and market requirements.

7   Monsanto L-dopa process

Probably the most elegant application of homogeneous catalysis is the synthesis of optically active organic compounds from non-chiral starting materials. This asymmetric induction can occur in many reactions catalysed by transition metal complexes, but the first commercial application was the Monsanto synthesis of the ℓ isomer of 3,4-dihydroxyphenylalanine, L-dopa, a drug used in the treatment of Parkinson's disease.[39–41]

The essential feature for selective synthesis of one optical isomer of a chiral substance is an asymmetric catalyst site that will bind a prochiral olefin preferentially in one conformation. This recognition of the preferred conformation can be accomplished by the use of a chiral ligand on the metal. The ligand creates what is effectively a 'chiral hole' within the coordination sphere of the metal. The key step in the L-dopa synthesis is the catalytic hydrogenation of the prochiral olefin, a substituted cinnamic acid (reaction (33)), the product of which is converted into the therapeutically effective isomer of dihydroxyphenylalanine in a subsequent step.

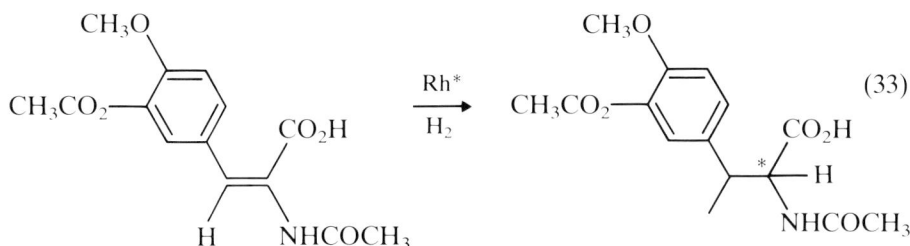

The catalyst is a derivative of the well known hydrogenation catalyst $Rh(PPh_3)_3Cl$, in which triphenylphosphine has been replaced by an asymmetric ligand such as the chelating diphosphine commonly known as DiPAMP (34). The catalyst precursor $[Rh(DiPAMP)(diene)]^+ BF_4^-$ is converted into a species of the type $[Rh(DiPAMP)H_2(ROH)_n]^+$ in aqueous ethanol or 2-propanol under the reaction conditions (3 atm hydrogen, 50°C).

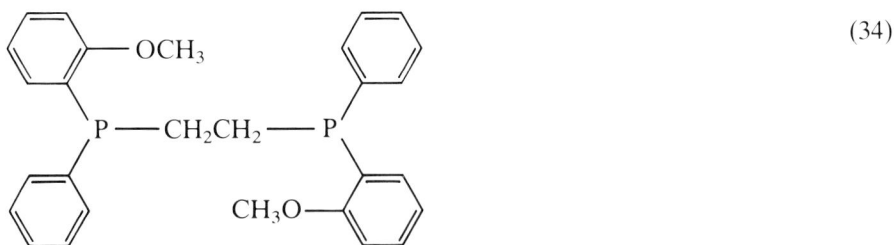

The substrate coordinates to the rhodium both through its carbon–carbon double bond and through the carbonyl function of the acetamide grouping. This rigid arrangement has the effect of constraining the substrate to present one face specifically to the rhodium atom. The transfer of hydrogen to the bound face of the olefin generates one optical isomer specifically. Optical yields, expressed as the enantiomeric excess of the desired conformation, are above 95%. The yields are very sensitive to solution pH and to the purity of hydrogen used. The L-dopa derivative is isolated by recrystallization from methanol, the catalyst being left in the mother liquor from which the rhodium can be extracted and isolated for re-use. It is reported that one pound of catalyst yields one ton of L-dopa. In systems of this type it can be envisaged that the catalysts are very expensive, not only because they contain rhodium, which can of course be recovered, but more importantly because they contain chiral phosphines which are not only difficult and expensive to synthesize but probably are also difficult to recover and re-use. However, in high added value situations such as these, high catalyst costs can clearly be tolerated.

8   Conclusions and future perspectives

From the preceding sections of this chapter it is clear that homogeneous catalysts play a very significant role in the processing of industrial chemicals. During the 1970s no less than five major new processes based on homogeneous catalysis were brought on stream and this progress has continued into the 1980s with the announcement and/or completion of a further two. Whether this impetus will be sustained in an era of more rapidly fluctuating economic conditions—recession, followed by only slow growth, etc—is extremely difficult to predict. The direction of future applications is certainly likely to change in concert with the trends towards the use of more energy-efficient processes and cheaper feedstocks, as discussed previously, and also with the increasing sophistication available in catalyst design. The latter should facilitate a shift in direction away from applications in large tonnage commodity chemicals towards much smaller tonnage 'high added value' or 'fine' chemicals such as that produced in the Monsanto L-dopa synthesis.

Some of the chemically most demanding transformations, namely the selective reduction of carbon monoxide to oxygenated products, the reduction of molecular nitrogen to ammonia and the activation of saturated hydrocarbons, have all been recently demonstrated to occur in the presence of soluble metal complexes—these were all previously believed to be the preserve of heterogeneous catalysts. It is likely that these areas will provide significant opportunities for the development of homogeneously-catalysed processes. For example, there has been considerable effort in the so-called area of C_1 chemistry, centred on the selective reduction of carbon monoxide in particular, but also on the transformation of simple molecules, e.g. CH_3OH, CH_2O, HCO_2H, CO_2 and CH_4 into higher carbon number products such as ethylene glycol, ethanol, higher alcohols and olefins. The driving force behind this effort has been to reduce the dependence on oil as the basic feedstock for petrochemical use and to replace it by a cheaper and, in terms of availability (via natural gas, heavy residual oil, shale oil or coal), more versatile feedstock such as synthesis gas.

The technology related to processes for the production of acetic acid and of acetic anhydride has already been described (sections 4.2 and 4.3). In addition, a particularly attractive target has been the direct conversion of carbon monoxide and hydrogen into ethylene glycol (reaction (35)), as a potential successor to ethylene oxidation technology.

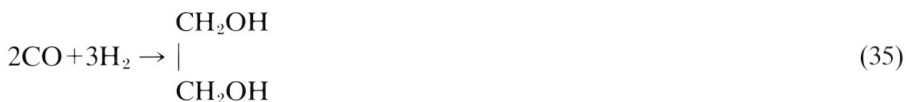

$$2CO + 3H_2 \rightarrow \begin{array}{c} CH_2OH \\ | \\ CH_2OH \end{array} \qquad (35)$$

Union Carbide have discovered, and worked extensively with, a homogeneous promoted rhodium catalyst which gives high selectivities to ethylene glycol and methanol with no methane formation.[42] Methane formation is a feature which is characteristic of almost all heterogeneous catalysts for carbon monoxide hydrogenation, the notable exception being the Cu/Zn methanol synthesis catalyst. Metals other than rhodium give considerably lower selectivities to ethylene glycol except in the case of a bimetallic promoted catalyst containing ruthenium and only minor amounts of rhodium[43] which, in acetic acid as solvent/co-reactant, yields the acetate esters of ethylene glycol and methanol in comparable selectivities and activities to the rhodium-containing catalysts described by Union Carbide. The problems with commercialization of both these systems are the requirements for very high operating pressures (750–1000 atm) to give good ethylene glycol/methanol selectivities (of 1.5–2.0:1) and the relatively low activities displayed by these catalysts. However, many leads with respect to the effect of promoters, solvents, etc., remain unexplored and the latter system at least is by no means optimized. Homogeneous ruthenium-based catalysts have been described for the production of C_1–C_3 alcohols and acetate esters directly from synthesis gas and their potential use as fuel additives discussed.[44] Because of the rapidly changing economic conditions—recession, reduction in oil prices, etc.—much of the world-wide research effort in this area has been reduced. But one thing is quite certain—the need for selective syntheses of higher carbon number molecules, particularly oxygenated products, from C_1 molecules will return and homogeneous catalysis has the potential to play a very significant role in this area.

Many opportunities exist for the application of homogeneous catalysts in the synthesis of fine chemicals. The use of asymmetric hydrogenation in the production of L-dopa, discussed in section 7, is only one example of many potential applications of asymmetric catalysis. In addition to asymmetric carbon–hydrogen bond formation the areas of asymmetric carbon–carbon bond formation and asymmetric carbon–heteroatom bond formation represent a fascinating array of opportunities.[45] For example, in the area of stereospecific oxidations, the kinetic resolution of racemic allylic alcohols by enantioselective epoxidation using a titanium tetraisopropoxide/tartrate reagent, in the presence of t-butylhydroperoxide as the source of oxygen, has led to a route to molecules of > 95% enantiomeric purity.[46,47] This approach has been utilized for the synthesis of (+)-disparlure (36), the sex attractant of the gypsy moth. In 1980 this was sold for

£1000 g^{-1} but using the asymmetric epoxidation route the current price is now said to be £100 g^{-1}.[45]

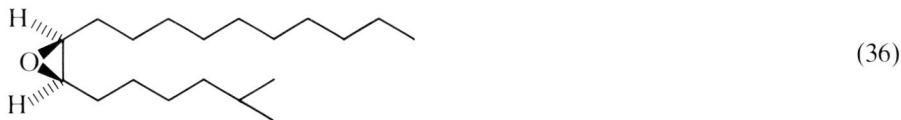

(36)

Clearly there are many other potential applications of homogeneous catalysis for chiral epoxidation and asymmetric synthesis in general, and this may prove to be the area of most significant growth in the future.

9   References

1   Bogdanovic, B., Spliethoff, B. & Wilke, G. *Angew. Chem. Internat. Edit.* 1980, **19**, 622.
2   Catalysis and organic synthesis. *Adv. Organometallic Chem.* 1979, **17**, and references therein.
3   Parshall, G.W. *Homogeneous Catalysis—The Applications and Chemistry of Catalysis by Soluble Transition Metal Complexes.* John Wiley, New York, 1980.
4   Masters, C. *Homogeneous Transition Metal Catalysis—A Gentle Art.* Chapman and Hall, London, 1981.
5   Parshall, G.W. *J. Mol. Catal.* 1978, **4**, 243.
6   Parshall, G.W. *Catal. Rev.—Sci Eng.* 1981, **23**, 107.
7   Collman, J.P. & Hegedus, L.S., *Principles and Applications of Organotransition Metal Chemistry*, University Science Books, Mill Valley, California, 1980.
8   *Catalysis and Chemical Processes.* Pearce, R. & Patterson, W.R. (Eds). Leonard Hill (Blackie), 1981.
9   Pino, P., Piacenti, F. & Bianchi, M. In *Organic Syntheses via Metal Carbonyls.* Wender, I. & Pino, P. (Eds), p. 43. Vol. 2, Wiley-Interscience, 1977.
10   Pruett, R.L., ref. 2, p. 1.
11   *New Syntheses with Carbon Monoxide.* Falbe, J. (Ed.) Springer-Verlag, Berlin, 1980.
12   Slaugh, L.H. & Mullineaux, R.D. *J. Organometallic Chem.* 1968, **13**, 469.
13   Tucci, E.R. *Ind. Eng. Chem. Prod. Res. Dev.* 1970, **9**, 516.
14   Fowler, R., Connor, H. & Baehl, R.A. *Hydrocarbon Processing,* 1976, **55**, September, 247.
15   Murrer, B.A. & Russell, M.J.H. In *Catalysis, Specialist Periodical Report,* The Royal Society of Chemistry, London, 1983, Vol. 6, p. 170.
16   Whyman, R. *J. Organometallic Chem.* 1975, **94**, 303.
17   Pino, P., Piacenti, F. & Bianchi, M. In *Organic Syntheses via Metal Carbonyls.* Wender, I & Pino, P. (Eds), p. 233. Vol. 2, Wiley-Interscience, New York, 1977.
18   *Eur. Chem. News.,* 1984, July 9, p. 34.
19   See, for example, Dubois, R.A., Garrou, P.E., Lavin, K.D. & Allcock, H.R. *Organometallics,* 1984, **3**, 649.
20   Philips, F.C. *Amer. Chem. J.* 1894, **16**, 255.
21   Smidt, J., Hafner, W., Jira, R., Sieber, R., Sedlmeier, J. & Sabel, A. *Angew. Chem. Internat. Edit.* 1962, **1**, 80.
22   Henry, P.M. *Adv. Organometallic Chem.* 1975, **13**, 363.
23   Hohenschutz, H., von Kutepow, N. & Himmele, W. *Hydrocarbon Processing,* 1966, **45**, November, 141.
24   Paulik, F.E. & Roth, J.F. *J.C.S. Chem. Comm.* 1968, 1578.
25   Forster, D., ref. 2, p. 255.
26   Robinson, K.K., Hershman, A., Craddock, J.H. & Roth, J.F. *J. Catal.* 1972, **27**, 389.

27   Forster, D. *J. Amer. Chem. Soc.* 1976, **98,** 846.
28   *Eur. Chem. News,* 1980, March 3, p. 24.
29   Schrod, M. & Luft, G. *Ind. Eng. Chem. Prod. Res. Dev.* 1981, **20,** 649.
30   Luft, G. & Schrod, M. *J. Molecular Catalysis,* 1983, **20,** 175.
31   Schrod, M., Luft, G. & Grobe, J. *J. Molecular Catalysis,* 1983, **22,** 169.
32   Ehrler, J.L. & Juran, B. *Hydrocarbon Processing,* 1982, **61,** February, 109.
33   *Chem. Eng. News,* 1971, April 26, p. 30.
34   Tolman, C.A., Seidel, W.C., Druliner, J.D. & Domaille, P.J. *Organometallics,* 1984, **3,** 33.
35   Druliner, J.D. *Organometallics,* 1984, **3,** 205.
36   Seidel, W.C. & Tolman, C.A. *Ann. N.Y. Acad. Sci.* 1983, **415,** 201.
37   Freitas, E.R. & Gum, C.R. *Chem. Eng. Progr.* 1979, January, 73.
38   Peuckert, M. & Keim, W. *Organometallics,* 1983, **2,** 594.
39   Knowles, W.S., Sabacky, M.J. & Vineyard, B.D. *Chem. Tech.* 1972, **2,** 590.
40   Vineyard, B.D., Knowles, W.S., Sabacky, M.J., Bachman, G.L. & Weinkauff, O.J. *J. Amer. Chem. Soc.* 1977, **99,** 5946.
41   Knowles, W.S. *Accounts Chem. Res.* 1983, **16,** 106.
42   Pruett, R.L. *Ann. N.Y. Acad. Sci.* 1977, **295,** 239.
43   Whyman, R. *J.C.S. Chem. Comm.* 1983, 1439.
44   Knifton, J.F., Grigsby, R.A. & Herbstman, S. *Hydrocarbon Processing,* 1984, January, 111.
45   Bosnich, B. *Chem. Brit.* 1984, September, 808.
46   Rossiter, B.E., Katsuki, T. & Sharpless, K.B. *J. Amer. Chem. Soc.* 1981, **103,** 464.
47   Martin, V.S., Woodward, S.S., Katsuki, T., Yamada, Y., Ikeda, M. & Sharpless, K.B. *J. Amer. Chem. Soc.* 1981, **103,** 6237.

Index